THE CHIEFS REMEMBER

THE CHIEFS REMEMBER

The Forest Service, 1952–2001

Harold K. Steen

Forest History Society, 2004

The Forest History Society is a nonprofit, educational institution dedicated to the advancement of historical understanding of human interaction with the forest environment. The society was established in 1946. Interpretations and conclusions in FHS publications are those of the authors; the Society takes responsibility for the selection of topics, the competence of the authors, and their freedom of inquiry.

Forest History Society
701 William Vickers Avenue
Durham, North Carolina 27701
(919) 682-9319
www.foresthistory.org

First edition

Design by Zubigraphics, Inc.

This publication was supported by the Lynn W. Day Endowment for Forest History Publications and supported with additional funds from the USDA Forest Service, New Century of Service, in recognition of the centennial anniversary of the USDA Forest Service.

Library of Congress Cataloging-in-Publication Data

Steen, Harold K.
The chiefs remember : the Forest Service, 1952–2001 / Harold K. Steen.--
1st ed.
p. cm.
Based on interviews with Forest Service chiefs.
Includes index.
ISBN 0-89030-063-1 (pbk. : alk. paper) -- ISBN 0-89030-064-X (hardcover :
alk. paper)
1. United States. Forest Service--Officials and employees--Biography. 2.
United States. Forest Service--History--20th century. I. Forest History
Society. II. Title.
SD127.S74 2004
634.9'092'273--dc22
2004025243

For Elwood R. "Woody" Maunder,
oral history pioneer and indefatigable practitioner

ON THE COVER: THE BADGE OF THE FOREST SERVICE

Identified with the Forest Service for twenty-five years as its printed emblem, its property-mark, the flag of its fleet, and, above all, as the service symbol and badge of authority worn by every officer, is the device of the shield-enclosed tree. Its history is less known, even by the thousands of wearers to whom it signifies so much and whose highest loyalty and effort it inspires.... When in 1905 the newly named Forest Service desired to supplant the circular nickeled badge that previously showed the authority of forest reserve officers, a designing contest was instituted at Washington, D.C. Gifford Pinchot, then Chief Forester, Overton W. Price and E. T. Allen comprised the judging committee and no rules of design were imposed.

The result was an interesting collection, some ingenious and artistic—scrolls, tree leaves and maple seeds—but there was much disregard of recognizable authority to enforce law or regulations. Not a single design satisfactorily combined essentials. So a new start, with specified requirements, was necessary. As a suggestion along this line, Allen, who as one of the judges was insistent upon a conventionalized shield of some kind to assure quick public recognition of authority and also suggest public defense as a forestry object, was tracing the Union Pacific Railroad shield emblem from a time folder which lay on his desk and inserted the letters "U. S." conspicuously. W. C. Hodge, now dead, who was watching him, suddenly sketched a conventional coniferous tree on a cigarette paper and laid it between the two letters to complete the symbolism. Another minute and "Forest Service" was written above and "U. S. Department of Agriculture" below. This three-minute combination of a railway folder and a cigarette paper satisfied all three judges so the contest was called off....

From The Shield and the Tree: Origin of the Emblem Which Has Been the Symbol of the Forest Service for Twenty-five Years. *American Forests and Forest Life, July 1930: 392.*

CONTENTS

FOREWORD

The Forest Service in an Age of Change

From its very inception, the Forest Service was no stranger to controversy. In the western states, many indigenous interests and their representatives in Congress were incensed by the withdrawals of vast areas of the public domain from private acquisition, development, and exploitation. These disgruntled groups exerted a great deal of pressure to mitigate the execution of the conservation policies, so the Forest Service was constantly embattled. When national forests were established in the East, the contention spread, and though it was less severe, the agency still had to deal with demands for more intensive use of the resources than was compatible with its mission. Moreover, several early chiefs strongly advocated federal regulation of logging on privately owned land, a stance that embroiled them in further discord. And other federal land management agencies with different priorities and philosophies and clienteles, such as the National Park Service, the Bureau of Land Management, and the Fish and Wildlife Service, were occasionally critics and rivals of the Forest Service. So it was never free of controversy.

Nonetheless, the Forest Service came to enjoy great prestige and support as the leading champion of natural resource conservation. Indeed, its unwavering advocacy and defense of its position in the face of heavy pressures and competition doubtless contributed to its standing, for its steadfastness was seen as devotion to the cause for which it had been created. But its status and popularity probably derived also from its reputation for administrative excellence. Its officers were dedicated professional foresters who, from top to bottom, subscribed to a common philosophy of resource management; their unity, teamwork, high morale, esprit de corps, integrity, and loyalty to the agency and its goals earned it widespread plaudits and respect, nowhere more than in Washington. Consequently, the Forest Service exercised great influence on conservation policy and achieved great autonomy in the executive branch, including a degree of independence sometimes taken amiss by some officials of the Department of Agriculture.

....

by Herbert Kaufman
Visiting Fellow, Department of Political Science, Yale University

As the testimonies of the chiefs in this volume attest, the agency's stature began to erode as it entered the second half of its first century of existence, not long after the end of World War II. Increasingly, it was not in the forefront of the course of events affecting it; rather, it was borne along by forces not of its own making. Its proactive stance became more reactive as other actors, both in and outside government, seized the initiative and changed the political landscape. The esteem it enjoyed in Congress, the environmental community, the press, and the public at large declined. A commission on the organization of the executive branch heard proposals to split up its functions and parcel them out to other agencies. New authority over the environment went mostly to newly created agencies. Even in its own ranks, discontent simmered.

What accounts for this turnabout? Why was an organization that had weathered so many challenges, resisted such powerful interests, contended effectively against strong competitors, and acquired so much experience now beset by these new problems?

Part of the answer may be that its world was changing too rapidly for it to keep up. In particular, its political base, the original Conservation Movement, underwent a dramatic transformation. At one time, in its struggles with interests whose overuse of resources could have desolated the national forests, the agency could count on the backing of the conservationists. After World War II, however, especially in the sixties and seventies, the environmentalists and the recreationists added their complaints about the agency to those of its traditional industrial adversaries. To be sure, the objectives of these two new critics were not entirely consistent with one another; the environmentalists called for more wilderness areas and sanctuaries for endangered species, from which all harvesting and development and even access roads and airplane landing strips would be excluded, while the recreationists hoped to limit logging and grazing but to open the gates wider to skiers, hunters, anglers, hikers, campers, and other recreational users and the developers who catered to them. Thus, while the two movements were not always in agreement with each other, they were both at odds with the long-established commercial users of the resources of the national forests. The political climate was more complicated than it had once been.

Confronted with the impossibility of fully gratifying all of these claimants simultaneously, the Forest Service could plausibly argue that it was steering an appropriate middle course. Nevertheless, it found itself in an unfamiliar predicament. Instead of facing its customary opponents with the support of its usual friends, it was now beleaguered from all sides by an unlikely (and shifting) conjunction of disparate detractors with incompatible agendas. And the new participants in the arena of natural resources policy making turned out to be politically adept, gaining allies among members of Congress and the executive branch and their staffs, enlisting the support of the media, bringing suit in the courts, pitting administrative agencies against one another, and taking part in the administrative proceedings of agencies at every opportunity. They could not be brushed off.

At the same time, the Forest Service found itself under attack for insufficient diversity in the composition of its staff. The environmental movement urged the appointment of personnel from disciplines and professions other than forestry, a position taken also by new members of its own ranks as the schools of forestry, from which the agency continued to recruit heavily, broadened their curricula (and even changed their names) to emphasize environmental sciences. Simultaneously, the civil rights movement surged, and the agency found itself under pressure to recruit more women and minorities.

So on every side, it was besieged. As the chiefs' oral histories confirm, many of the agency's responses were thrust upon it by other potent operators—Congress exercising its legislative, appropriations, and oversight powers; political executives employing their powers of appointment, command, and budgetary control; courts issuing their interpretations of the law; interest groups mobilizing to lobby and agitate; and other administrative agencies, now invested with authority overlapping and competing with the Forest Service's mandate, setting the pace. It was therefore perceived as not leading the way but being dragged along reluctantly. One would think an organization as experienced, informed, sophisticated, and skillful as the Forest Service would have been able to adapt easily. Why was it slow to respond?

Ironically, the practices that had made it so successful in the past now made adjustment difficult. To ensure that its field officers—scattered all over the country, charged with a broad variety of responsibilities, working under exceedingly varied conditions and in many types of communities, and to whom it was therefore necessary to delegate substantial authority—would handle their duties as their superiors wished, the agency had instituted extensive controls on field behavior, including an elaborate set of substantive, financial, and procedural directives covering a host of contingencies, and introduced a well-developed feedback system to detect and discourage deviation from the directives. Equally important, if not more so, it had adopted measures to instill in its personnel the agency's perceptions, values, and outlook so that they would, of their own volition, behave as the leaders wished, or as the leaders themselves would behave if they were in the shoes of their subordinates. The agency's criteria of decision making and action became the personal, internalized criteria of the individuals within it. And since higher positions were normally filled by promotion from within, these criteria permeated the organization from top to bottom. Although all organizations indoctrinate their personnel and strive to maintain stability, the Forest Service was distinctive because it did these things exceptionally well—to its great advantage—for a long time. It created a process of total acculturation that turned out to profoundly affect the substance of agency policy.

Without that process, the widely dispersed field forces might have fragmented into separate, autonomous fiefdoms. Each field unit, immersed as it was in its own distinctive aggregation of resource characteristics and local relationships and perceptions of its needs and opportunities, might have gone its own way. But the modus operandi of the Forest Service kept these powerful centrifugal forces in check and molded its entire workforce into a unified body faithfully carrying out the policies enunciated by its leaders. In addition, it engendered strong ties among the members, whose pride in the organization and its mission was palpable and whose high morale and vigorous spirit were much admired. The integrative mechanisms served the agency well when its goals were not disputed within the conservationist community and when its practices were not aggressively challenged by civil rights activists.

When the new environmentalists and the proponents of minority rights flexed their muscles in the sixties and seventies, however, their programs lay outside the prevailing homogeneous outlook in the agency, and they were regarded by many in the Forest Service as extremists or even as cranks. "Multiple use" had been a cornerstone of agency policy from the start, and in the eyes of the organization its definition of the term struck a reasonable balance among the diverse and often competing demands upon the resources for which it was responsible. In its view, each special interest sought undue emphasis on limited concerns, frequently to the detriment of other uses. It was determined not to favor one activity over others, and it denied the charge that it was in thrall to the logging and grazing interests. And with regard to civil rights, the agency's employment practices were, alas, in keeping with the norms throughout the country; it did not construe its mission as requiring it to embark on a crusade for social justice that might jeopardize its effectiveness. Forest Service officials, socialized to the organization's established perceptions and beliefs, were convinced they were on the right course. Critics called their uncompromising resolve self-righteous and contemptuous of other opinions; they saw themselves as principled and committed to the public interest. That's why, when the new forces managed, in a relatively short period, to convince influential legislators and political executives and judges and journalists of the validity of their causes, the Forest Service could not bring itself to abandon its well-established, deeply ingrained values and patterns of behavior.

Unquestionably, however, the agency has been yielding to the winds of change. Many professional specialties are now represented at every level; it is no longer made up almost exclusively of foresters, and some of the recent chiefs themselves were not even foresters by profession. More women and ethnic minority members have appeared on its rosters and in higher positions of authority. It has expanded—and it energetically guards—its wilderness areas. It has taken vigorous measures to protect endangered species and scenic sites. Recreation now has a higher place on its agenda. Diverse environmental values receive more emphasis. Fear of timber famine no longer dominates its approach because new technologies of wood use and the globalization of timber markets have eased shortages. As is evident from the comments of the chiefs in this volume, the Forest Service has truly broadened its perspectives.

It therefore is, and will doubtless continue to be, a major element in the public management of natural resources, and students of this subject will go on studying it attentively. And so will students of organizational behavior and public policy, for its transformation, whatever the reasons for it, raises another set of interesting questions: Will the new diversity in its composition and the realignment of its priorities weaken the unity that its past practices produced? Will internal harmony and cohesion give way to dissension? Under these new conditions, will it still be able to ensure that what happens on the ground conforms to the wishes and expectations of its leaders? If not, will it be able to regain the stature and the influence on public policy it enjoyed earlier? Can it balance the benefits and costs of unity and diversity in such a fashion as to avoid endlessly oscillating between them, and instead achieve a durable stability?

As the Forest Service embarks on its second century, it will no doubt reach new heights of excellence and encounter new hazards, in the course of which it will provide us with answers to some of the questions posed here—and possibly inspire a new set of questions as well. And five decades hence, I expect a new compilation of the reminiscences of the chiefs in office between now and then will prove as interesting and instructive and stimulating as the present compilation is to this generation.

INTRODUCTION

"Watch the bureaus, not the cabinet...The new [Reagan] administration's real tone will be determined less by those at the top than by the second line." This quote is from *The Administrative Behavior of Federal Bureau Chiefs*, by political scientist Herbert Kaufman. For this 1981 study, Kaufman looked closely at six rather different federal bureaus: the Social Security Administration, the Internal Revenue Service, the Customs Service, the Food and Drug Administration, the Animal and Plant Health Inspection Service, and the Forest Service. Kaufman explained, "I decided to include the Forest Service because I had previously examined its field officers in great detail [in *The Forest Ranger: A Study in Administrative Management*, 1960] and was therefore familiar with much of its history and operations."

For a year, he observed these six agency heads "as they did their jobs. By watching them at work, literally spending whole days looking over their shoulders," he found that the "willingness of all the chiefs...to participate in the research is rather remarkable in the annals of government bureaucracies. ...the relative ease of access to the inner working of the federal administrative establishment must be extraordinary compared with the officiousness and defensiveness of administrative officers and agencies in most parts of the world."

Kaufman concluded that to be effective, bureau chiefs needed to have four basic qualities. The first quality was "a juggler's disposition." Many people, such as scientists, are able to excel by focusing on a single topic for a period of time. Chiefs, however, live in a "simultaneous mode," with many things coming at them all at once, "not in single file. Their days are splintered." A chief could spend as much time on the problems of a single employee as on a national policy.

A chief must have "patience." A chief's time is not his own. "While he can bargain a little over the precise timing of meetings and hearings requested by his superiors or Congress or even by his staff, his schedule is shaped in large part by their demands and needs, and by outside groups as well." Kaufman noted that it "takes special patience to show deference to people with power who do not inspire admiration."

"Self-control" is also essential. "These positions call for still another form of tolerance—the capacity to resist the impulse to meddle with things that are going reasonably well." An overaggressive administrator "runs the risk of disrupting smooth-running routines, stirring up antagonism, and generating anxiety without producing enough benefits to justify the turmoil."

Finally, a chief must have "interrogatory skill" because he spends "most of his time receiving and reviewing information. To do so effectively calls for adroitness in putting questions, assessing replies, discerning gaps and inconsistencies, distinguishing soft from hard evidence, and sensing what has been withheld or designed to deflect a line of inquiry." Kaufman also pointed out that "sharp questioning demands personal security because a chief who engages in it frequently exposes his limitations." Specialized subordinates generally have superior expertise in their fields. "A chief may sometimes be the only one in the room having difficulty with a concept or an argument or a position that the technicians seem to grasp easily and agree upon." A chief who forgoes "challenging interrogation to avoid exposing his shortcomings...pays a high cost for his insecurity."

The narrative that follows presents excerpts from a series of interviews with Forest Service chiefs who generally demonstrated that they possessed Kaufman's four qualities. John McGuire was chief at the time of his study, but the context applies to all.

Richard E. McArdle became Forest Service chief in 1952, during Harry Truman's presidency; Edward P. Cliff succeeded McArdle in 1962, John R. McGuire followed Cliff in 1972, R. Max Peterson in 1979, F. Dale Robertson in 1987, Jack Ward Thomas in 1993, and Michael P. Dombeck in 1997. Dombeck resigned in 2001 at the beginning of George W. Bush's presidency. Their tenures as chief thus span fifty years, a half-century of rapid change and increasing controversy. It is history's good fortune that these seven conservation leaders agreed to be extensively interviewed, adding greatly to our understanding of the issues and challenges. This book centers on these interviews.

The interviews capture the recollections of seven accomplished individuals who have different personalities and respond to similar questions not necessarily in similar ways. Too, four people conducted the seven interviews, each asking questions in his own fashion and editing the transcript according to personal views of just what the final copy should look like. The questions themselves show parallels but also reflect changing times, changing controversies, and changing values; the questions asked McArdle are rather different from those posed to Dombeck. The narrative in this book, whether paraphrases or direct quotes, reflects what each said about a topic at the time of the interview; the same question asked in a different context on another day may well have yielded a different answer. For the sake of consistency, stylistic differences between the several transcripts have been silently edited.

In the pages that follow, the former chiefs look back at the issues faced during their administrations, as they saw them. The vast majority of Forest Service decisions and actions take place below the chief's level; the interviews tell us what chiefs think about and what they do. There is no attempt here to be complete or comprehensive; instead, this is a fair sample of each tenure. Too, the narrative is present-minded, a historian's term that means the selected topics are those seen as important today, which at times causes a distortion of "pure" history.

The first multiple-use and wilderness bills were introduced in Congress in 1956 while McArdle was chief; the Multiple Use–Sustained Yield Act became law in 1960, and the more controversial wilderness bill was debated four additional years and passed in 1964. It is fair to say that Multiple Use–Sustained Yield ratified Forest Service management practices that had evolved to 1960, but the Wilderness Act can be seen as only the beginning of congressional prescriptions that reduced the agency's management options. Both wilderness and multiple use remain major issues today. Also an issue is just how much timber should be sold from the national forests, and McArdle narrates his proposal to increase the cut to twenty billion board feet by the year 2000, a proposal that was adopted by both the outgoing Eisenhower administration and the incoming Kennedy administration. By the 1990s, such cuts were politically inconceivable.

Rachel Carson's *Silent Spring* appeared in 1962, the same year that Cliff became chief. Publication of *Silent Spring* with its clarion call for safer pesticide use provides a convenient—if arbitrary—beginning point for the environmental movement. The Forest Service and all other agencies and institutions during the 1960s struggled to cope with an avalanche of changing values that included civil rights and its closely related mandates for nondiscriminatory hiring practices. During his final year as chief, Cliff observed that the Forest Service had been poorly prepared to deal with the suddenly new world in which it operated.

On New Year's Day 1970, President Richard Nixon announced to the nation that he had signed the National Environmental Policy Act and that the 1970s would be the "environmental decade." NEPA authorized creation of the Council on Environmental Quality, required preparation of environmental impact statements that included public involvement in projected activities, and mandated an "interdisciplinary approach" to planning. Nixon then used presidential authority to create the Environmental Protection Agency to be the nation's "cop" for enforcing the ever-lengthening list of environmental laws. It would be some time before the Forest Service and other agencies realized the full significance of NEPA and EPA.

When McGuire became chief in 1972, wilderness and clearcutting were vexing issues, coupled with a growing workload caused by NEPA-related litigation. Courts frequently agreed with plaintiffs that the Forest Service's environmental impact statements, including those prepared for proposed wilderness areas, were "inadequate." Clearcutting, long a controversial practice, was attacked with increasing success; a court decision in 1973 declared that clearcutting violated the 1897 Organic Act, the cornerstone of agency policy until supplemented by the 1960 Multiple Use–Sustained Yield Act. The judge also admonished the agency to seek to change the law, if clearcutting was desirable, rather than dispute his decision. The 1976 National Forest Management Act, which permitted clearcutting on national forests under prescribed conditions and substantially revamped the planning process, was a result. For all too brief a period, it seemed that the Forest Service and its critics had found common ground.

Shortly after Peterson became chief, Ronald Reagan succeeded Jimmy Carter as president. The immensely popular Reagan had promised to reduce the size of the federal government, to restrict federal involvement in local issues, and to increase commodity use of public lands. The increase in resource use included moving the allowable cut beyond McArdle's earlier target of twenty billion board feet. Thus, Peterson faced the task of reducing the Forest Service workforce by twenty-five percent while the agency was preparing detailed plans for more timber sales and aggressively recruiting women, minorities, and specialists in nonforestry disciplines. Alarmed by Reagan administration policies, environmental groups mounted hugely successful membership drives and implemented a broad range of strategies to safeguard and extend their past successes. Increasingly, they used endangered species as a lever, citing the 1973 Endangered Species Act but mainly the diversity section of the National Forest Management Act. The northern spotted owl would shortly become an icon of environmental concern.

Robertson had been Peterson's associate chief, so he was familiar with the issues at the chief's level. By now the spotted owl—jobs versus owls—controversy was nearing the boiling point, and the pressures on the chief were great indeed; President George H. W. Bush's chief of staff even directed that he be fired. In the midst of all this furor, Robertson came to realize that traditional forestry practices had "hit the wall" and that multiple-use management of the national forests had created endangered species. Material changes were in order; the incremental policy shifts of the past were no longer adequate. Ecosystem management—an overlay for multiple use that required a much broader context—became official Forest Service policy. Robertson was removed from office during the early months of the Bill Clinton administration, but ecosystem management would continue to guide the agency.

Thomas was the lead scientist of the committee charged with studying the spotted owl and developing options for protecting the species while also allowing a degree of traditional forestry practices to continue. The Clinton administration named him as Robertson's successor, a controversial selection in that he was not a member of the Senior Executive Service and thus was ineligible to head an agency except via a direct presidential appointment. The nature of his appointment troubled Thomas, as did the blatant politics of day-to-day agency life in Washington, D.C.

Thomas gave priority to a fuller definition and adoption of ecosystem management. He also addressed increasingly uncertain lines of authority, including direct White House intervention in field-level decisions, and overlapping and conflicting legislative mandates. Closure of an Alaska pulp mill, protection of Columbia River salmon, and an unusual number of firefighting fatalities also demanded much of the chief's time.

Dombeck, a fisheries scientist for the Forest Service, was invited to transfer to the Bureau of Land Management while Robertson was chief. Later on he was appointed acting BLM director and worked closely with Chief Thomas on shared issues, such as fire suppression and protection of endangered species. During his final months at the Forest Service, Thomas had repeatedly recommended to the secretary of Agriculture that Dombeck be the next chief, and the secretary, who had also interviewed other candidates, agreed.

Dombeck placed a moratorium on entering fifty-eight million acres of roadless areas for commercial purposes, which was made official by presidential order. He worked to build better ties to the Department of Agriculture, the White House, and Congress. Favorable reaction to those better ties included the largest budget increase in the agency's history. His three-year tenure as acting director of the Bureau of Land Management makes him uniquely qualified to compare Forest Service and BLM cultures, and the contrast he draws will surprise many readers.

Many thanks are due first and foremost to the seven chiefs who gave freely of their time to be interviewed. Thanks, too, to Elwood Maunder, who interviewed McArdle, and to David Clary and Ronald Hartzer, who interviewed Cliff. National Park Service historian Richard Sellars, Grey Towers National Landmark Director Ed Brannon, Forest History Society President Steve Anderson, and Forest Service Assistant Director for Ecosystem Management Coordination Sharon Friedman read the manuscript and offered many suggestions to improve the text. Linda Feldman, as coordinator for the 2005 Forest Service centennial, provided financial support, and Forest Service historian Jerry Williams offered congenial advice. Chiefs Max Peterson, Dale Robertson, Jack Thomas, Mike Dombeck, Forest Service visual information specialist Renee Green-Smith, and the U.S. Geological Survey provided photographs, as did the Forest History Society. Marjory McGuire provided the Wendelin cartoon. Also at the Forest History Society Cheryl Oakes and Michele Justice helped in all the ways that they typically do. And finally, this is the third book of mine that Sally Atwater has edited; it's amazing what fresh eyes and a lot of talent can do to improve a manuscript.

CHIEF RICHARD E. McARDLE

USDA FOREST SERVICE PHOTO

In 1899, the year Richard E. McArdle was born, the future Forest Service was still an infant itself, known as the Division of Forestry, and the national forests—then called the forest reserves—comprised only some forty-six million acres, all in the West. Gifford Pinchot had spent but a year in his position as chief forester, hiring assistants, building his staff, and planning the acquisition of more land. Forestry was a young profession in the United States, and its Progressive promise—using scientific knowledge to achieve both conservation and material improvement for the common man—would capture the interest of McArdle. He left his home in Lexington, Kentucky, to earn his bachelor's and master's degrees in forestry from the University of Michigan, and then began working for the Pacific Northwest Forest and Range Experiment Station in 1924. On leave from the station, he earned a Ph.D., also from Michigan, in 1930.

Back in Portland from Ann Arbor, McArdle continued his silviculture research but with special interest in fire. He had an inventor's eye and clever hands, and station records include his sketches and specifications for handmade fan psychrometers to measure relative humidity in the field and tinted glasses to help fire lookouts see better though smoke and haze. No doubt his most important

Source: Dr. Richard E. McArdle: An Interview with the Former Chief, *by Elwood R. Maunder. Durham, North Carolina: Forest History Society, 1975. McArdle was interviewed in 1975.*

publication, authored with Walter H. Meyer, was *The Yield of Douglas Fir in the Pacific Northwest*, USDA Technical Bulletin 201, in 1930. Bulletin 201, as it was generally called, quickly became the bible in the Douglas-fir region and was much used for decades by field practitioners as well as forestry students. Bulletin 201 was centered on the now-obsolete concept of a "normal stand," whereby a forester could predict the future of a young stand by comparing it with tables in McArdle's work. In other regions of the United States, Forest Service scientists produced similar volumes for all commercially important tree species. Because they could now forecast yield, managers were more willing to make long-term investments in forest management.

His affability and his penchant for practical jokes became legendary. It would be a badge of honor at a meeting to have McArdle, seemingly without effort, one-handedly loop a string puzzle though your lapel buttonhole. Try as you might, it was impossible to remove without cutting the string (even though McArdle could). But you wouldn't try too hard anyway; as would others, you would wear it proudly through the day. McArdle could remember your name, years later and apparently without hesitation. Thus, he could work a room and make everyone glad that the chief was there. During his interview, McArdle admitted that he would study the list of attendees with care ahead of time to refresh his memory. His charming manner was such that while he was chief, even the harshest critics of his policies could not help but like him.

In 1934, McArdle accepted the deanship at the University of Idaho School of Forestry. After a year in Moscow, he returned to the Forest Service as director of the Rocky Mountain Forest and Range Experiment Station in Fort Collins, Colorado. Three years later he moved to Asheville, North Carolina, as director of the Appalachian Forest Experiment Station. By 1944, he had moved to Washington, D.C., as assistant chief (the position of deputy chief would not be established until 1963) for State and Private Forestry. During the final year of Harry Truman's presidency, on July 1, 1952, McArdle was appointed chief.

REGULATION OF INDUSTRIAL LAND

As it is today, the Forest Service in 1952 was deep in controversy, but the issue was very different. In 1919, Gifford Pinchot had begun a vigorous campaign for federal legislation to give the Forest Service regulatory authority over logging practices on privately owned land. Off and on for the next three decades, language as harsh as any heard today from environmentalists was put forth by leaders of the lumber and livestock industries, as they and their congressional supporters successfully fended off federal regulation.

Chief McArdle with staff of the Coeur d'Alene National Forest, 1960. McArdle's ability to remember employees' names was an agency legend. USDA Forest Service Photo.

Chiefs Ferdinand A. Silcox (1933–1939), Earle H. Clapp (1939–1943), and Lyle F. Watts (1943–1952) had been especially aggressive in pursuit of regulatory authorities, and as Watts's successor, McArdle inherited the conflict. During his eight years as assistant chief, he had been directly involved with industry activities. In his opinion, the regulation controversy had been deeply and nonproductively divisive among those with a common interest in proper management of forest and range lands. Although he was not "completely antiregulation," he decided that it should be given low priority. Other legislative needs should be pursued instead.

One of his first acts was to recommend to the secretary of Agriculture that Edward P. Cliff be promoted to assistant chief in charge of National Forest Resource Management. Later, Cliff would remember that Chief Watts had been "closely identified" with the New Deal of Roosevelt and Truman. Watts believed that he could not "survive a transition" from a Democratic administration to a Republican and therefore elected to retire to allow his successor to be firmly in place before the 1952 election. During the presidential campaign, Watts was active in Conservationists for Stevenson, supporting Democratic candidate Adlai Stevenson, who twice lost to Dwight Eisenhower.

Cliff recalled that McArdle did not make a public statement about ending Forest Service regulation efforts. He believed that he could not immediately repudiate a major, "perhaps *the* major, publicly expressed policy of the past three chiefs." Although records of the Eisenhower administration suggest otherwise, McArdle clearly stated in his interview that "none of my new political bosses ever talked to me about regulation..." He had made his decision "months before" Eisenhower was inaugurated, and "I wasn't much interested" in what the new people thought about it. But the damage had already been done, and another generation of public and private forestry leaders would remember the regulation issue and be wary of each other. Thus, even as both the Forest Service and the lumber and range industries were being more and more tarred by the same environmentalist brush, these traditional adversaries could not bring themselves to sit down and fashion a mutual response. As it turned out, during the 1970s, as Congress passed regulatory laws, it would be the Environmental Protection Agency, the Army Corps of Engineers, and the Fish and Wildlife Service—not the Forest Service—that would be given authority.

RECREATION AND WILDERNESS

As part of his historical review of the regulation issue, McArdle also summarized a slowly accelerating shift of values by noting the increase in interest in preservation of scenery instead of preserving timber supplies. Proposals for regulation of logging practices had been intended to protect timber supplies by reducing waste and fire hazard and increasing reforestation. Such regulation would cover only private lands; after all, until 1940, ninety-eight percent of timber cut in the United States came from private holdings. It is not surprising, then, that the general public had come to see the national forests as places where trees were preserved, a notion much bolstered by blitzes of Forest Service regulation propaganda that portrayed the logging of private lands as "devastation."

By the 1950s, however, Forest Service timber harvest was spiking toward a full third of the national cut. To those not trained in forestry, logging on national forests looked pretty much like logging on private lands, and they did not like what they saw. As the line on Forest Service graphs reporting the timber cut arced upward, so did membership in organizations that supported preservation of scenery. The conflict would go public during the 1960s, and those who saw

Chief McArdle with President Dwight Eisenhower and Montana Governor John Hugo at the Aerial Fire Depot, Missoula, 1954. USDA Forest Service Photo.

the national forests mainly in terms of timber supply began to lose influence that would never be regained.

Recreation on the national forests—camping, hunting, fishing, and vacationing at summer homes and resorts—had always been significant. With the advent of the automobile and the concurrent extension of the public road system, recreational visits to the forests boomed. In 1924, President Calvin Coolidge spoke of the need for a definite national policy on outdoor recreation, and shortly thereafter 309 delegates from 128 organizations attended the National Conference on Outdoor Recreation in Washington, D.C. That same year, the Forest Service set aside the nation's first wilderness area on the Gila National Forest in New Mexico. Recreation now had its own box in the agency's organizational chart, and for a period during the 1930s, wilderness advocate Bob Marshall headed the program. Recreation and wildlife were administratively paired—back then, wildlife was seen mainly as game, and hunting was a form of recreation—and continued to expand their influence on Forest Service programs.

In 1958, McArdle watched as Congress passed the Outdoor Recreation Resources Review Act, which authorized creation of a commission made up of members of Congress and presidential appointees. The Forest Service and the National Park Service at that time were toe-to-toe in competition over recreational leadership. In 1956, the Park Service had launched its Mission 66, which aimed to restore and improve the parks to their fullest potential in ten years. The Forest Service quickly countered with Operation Outdoors, which included creation of a recreation program in its research branch, plus the upgrading and new construction of much-needed recreation facilities. McArdle hoped that the new commission would be "competent," so that its recommendations would be "more valuable." As it turned out, McArdle played a direct role in the selection of the seven presidential appointees to the new commission. Agency leaders had already created a list of the sort of mix that they believed was needed—representatives of the conservation community, the states, the forest industry, and the academy.

Sherman Adams, Eisenhower's chief of staff, was former governor of New Hampshire and had earlier been in the lumber business, and he knew McArdle by name. Too, he had written an antiregulation article that had pretty much set the official tone of the incoming administration. Adams's telephone style was always marked by abruptness, and shortly after passage of the Recreation Commission Act, he called McArdle. The chief recalled that he did not even identify himself: "I've got [Secretary of the Interior] Fred Seaton sitting here beside me. I've told him I am not going to pass along to the president any recommendations for membership of this recreation review commission that don't have your [meaning my] approval." Adams continued, "I want you to go over to Fred's office right away and you two sit down and reach agreement on who the president should appoint."

With that mandate, McArdle went to Seaton's office, where he thought the opening greeting was "a bit on the querulous side." The secretary stated that he should be allowed to name at least one appointee. McArdle agreed and asked who he had in mind. Seaton said M. Frederick Smith, who McArdle pointed out was a registered lobbyist. The surprised secretary said that he would have Smith's name removed from the list. Then McArdle said he believed the president was already committed to appoint Laurance Rockefeller as commission chairman, and they left that slot alone. As their meeting progressed, it became obvious to both that Seaton had not done his homework and that the Forest

Service list was superior. The final score for nominees was President Eisenhower 1, Chief McArdle 5, Secretary Seaton 1. McArdle thought that to be a "pretty fair split." Their final decision was to select an executive director for the commission, and Seaton had a name ready. It turned out that the secretary's preference just happened to be McArdle's brother-in-law. Learning that, Seaton said, "We don't want him." Later, as McArdle delightedly recounted, the secretary explained to a large audience that the brother-in-law would have headed the commission, except for McArdle's "strong objections."

The commission's report was in fact voluminous, well done, and influential. Its charge was to inventory and evaluate recreation resources and needs, projected to the year 2000. The report predicted a threefold increase in recreational demand by the end of the century. One of its recommendations that was adopted during the Kennedy administration was the creation of a Bureau of Outdoor Recreation. McArdle's brilliant but acerbic assistant chief for legislative affairs, Edward C. Crafts, left the Forest Service to be its first director. Since the new agency had no real constituency, its fortunes eroded and eventually it went out of existence.

MULTIPLE USE–SUSTAINED YIELD ACT

The ever-larger shadow cast by recreation convinced Forest Service leaders that the agency needed legislative authorization to establish some sort of balance of programs. Most of its essential authority stemmed from the 1897 Organic Act, which stated that no national forest could be created except to protect water and timber supply. The language clearly reflected congressional intent that the authority it had given the president to establish forest reserves (since 1907 called national forests) by proclamation was not to create national parks. The reserved land was to contribute significantly to the West's growing economy. Other uses, such as grazing, were officially added over time, and Congress recognized them via appropriations measures, but there was no specific statutory language to cover the full range of Forest Service programs that had evolved by the 1950s. McArdle assigned Crafts the task of drafting language, and in 1956, Senator Hubert H. Humphrey introduced a multiple-use bill. McArdle gives full credit to Assistant Chief Crafts for his legislative skills in seeing the bill through to passage in 1960.

The forest products industry and the Sierra Club, no surprise here, each had concerns about the bill's language. Industrial support was gained by adding language that the new law would be "supplemental to, not in derogation of" the 1897 Organic Act, supposedly thus ensuring that national forest timber would

Chief McArdle (right) presenting *Timber Resources for America's Future* to Secretary of Agriculture Ezra Taft Benson, with Assistant Chief Crafts looking on, 1958. The timber resources review announced the end of Pinchot's "timber famine." USDA Forest Service Photo.

continue to be sold. Although never agreeing to support the act, the Sierra Club withdrew its objections after language was added to say wilderness was consistent with multiple use. Too, the environmentalists won a minor victory during markup when "recreation" was called "outdoor recreation," thus putting their primary interest at the alphabetical head of the list of uses: "outdoor recreation, range, timber, watershed, and wildlife and fish purposes." If "fish and wildlife" had been used instead of "wildlife and fish," then that use would have been first. The states threw in their support when assured that their traditional control of hunting would not be affected; likewise, the mining industry saw language that protected its interests. Ducks now in a row, the bill passed without much additional controversy. Senator Humphrey's wilderness bill, which had also been introduced in 1956, proved more controversial and required another four years of debate before passage.

In 1960, the Fifth World Forestry Congress convened in Seattle, with McArdle serving as its president. The theme of the congress was multiple use, providing an excellent platform for the Forest Service to spotlight its new authority. In

Assistant Chief Edward C. Crafts

Multiple Use, a Milestone

The Multiple Use–Sustained Yield Act came partly because of the continual pressures of the timber industry to cut more timber. It came partly because of more intensive use of the national forests for water, for recreation, and as a matter of fact, for all of the national forest resources. We anticipated that unless we had statutory protection looking ahead to the future, the Forest Service would be in difficulty on two counts.

First, we should have a statutory prohibition to prevent overuse of any national forest resource, and this would give us a legal prop to fall back on when the pressures got to be overwhelming to overcut the timber. Second, we felt that all the resources of the national forests should be treated with equal attention and equal priority.

We felt sure that the time would come as the country and the economy grew in complexity, when we would not be able administratively to withstand the pressures to overcut, overgraze, and overrecreate on the national forests. Also the uses in many instances were competitive, not necessarily complementary, and the time would come when one use would tend to be predominant over the others. This we felt would be inequitable to the public at large.

Therefore, we decided, after most careful consideration and much internal reluctance inside the Forest Service, to propose the Multiple Use–Sustained Yield Act. Now it is usually referred to as the Multiple Use Act, which is too bad because it overlooks a very key part. This was sustained yield, which means continuous production at high levels, but not overuse. Sustained yield is just as important as multiple use. Multiple use assures that all of the resources will be considered, but sustained-yield assures that none of them will be overutilized.

We were hesitant to try for the act. We knew it was a gamble; we just didn't know whether we could get it. Many of the timber people felt that they then occupied a preferential position over other resources and they wanted to continue to be first. So initially, they opposed the legislation. They didn't want to recognize that grazing and recreation should have equal consideration.

There was great uncertainty in the Forest Service whether we would be successful in its passage. If we tried and failed, then we would be faced with adverse legislative history, which would be very bad. So we entered this knowing the risks involved, but determined to make it work if we could.

I might say that two of the real leaders of the Forest Service were strongly against our trying it, according to my recollection. One was [Assistant Chief] Chris Granger, who had recently retired but nevertheless was still brought in frequently on major policy questions. The other was Ed Cliff, who at the time was assistant chief in charge of National Forest Resource Management. McArdle and I and the

continued on next page

continued from previous page

rest of the staff, however, thought that it was then or never, so we decided to take a shot at it. It turned out to be the hardest job of getting a piece of legislation enacted that I faced during my period in the Forest Service.

I truly believe that this was the major legislative accomplishment during that decade, and I think the Forest Service generally has since come to this conclusion. They constantly refer to it now as the basis for their policy, and it ranks along with the 1897 [Organic] Act, the 1905 [Transfer] Act, the Weeks Act of 1911, and the Clarke-McNary Act [of 1924], as one of the real milestones of Forest Service basic legislation.

...Among the main contributors were certain leaders in Congress, as always, and Bernie Orell, vice president of Weyerhaeuser who almost single-handedly brought the timber industry along. Orell pretty much laid his own reputation on the line with the industry to do so. The Forest Service should be forever grateful to him. Without Orell's foresight, this act would never have been passed, and he should be given great credit for what happened.

Orell could see the handwriting on the wall. Also, as an ex-state forester he basically believed in sustained yield. He thought that if the timber industry opposed a proposal like this it would put them in a very disadvantageous public light, much as the stockmen were placed in 1953 [when they unsuccessfully supported a grazing bill]. The timber industry would be pictured to the public as being against sustained yield.

He also realized that other uses were growing in importance and had to receive their equitable share of attention. I would say that it was a real statesmanlike position on the part of Orell. It certainly was not a self-serving position as far as he was concerned. He had enough stature to get the industry to go along and to call off their dogs, but he had a difficult time doing this. He staked his own reputation and succeeded.

Clint Anderson was probably the leader [in Congress]. He was just a tower of strength and a real statesman. His health was still good at that time. He was still close to the Forest Service, remembering his days as secretary of Agriculture. He helped us immensely.

...That is about all I can say about the Multiple Use–Sustained Yield Act, except to reemphasize that legislatively this was the major affirmative Forest Service accomplishment during that decade. The Multiple Use Mining Act was probably second. The major negative accomplishment was prevention of passage of the Stockmen's Grazing Bill.

From Edward C. Crafts, "Congress and the Forest Service, 1950–1962," typescript of an interview conducted in 1965 by Amelia R. Fry. Berkeley: Regional Oral History Office, The Bancroft Library, University of California, 1975. Courtesy of the Bancroft Library.

his presidential address, "The Concept of Multiple Use of Forest and Associated Lands: Its Values and Limitations," McArdle explained to delegates from fifty-five nations that "multiple use is a familiar term to foresters of the United States." He described its evolution in the United States and listed conflicts and problems. He stated that "the five basic renewable resources shall be utilized in the combination that will best serve the people. Emphasis is on utilization, not preservation."

He also listed the uses, but in reverse order, making recreation last instead of first. An inadvertent slight or not, it fit well with a series of advocacy skirmishes at the fringes of activity. Sierra Club representatives with no official spot on the program talked of "multiple abuse." Four decades later the flap seems rather tame, but this sort of "ungentlemanly" and even "shocking" behavior only hinted at what was to come, as the Forest Service lost more and more of its long-held support from the Sierra Club and other environmental groups. For example, since the 1920s the club had routinely elected the Forest Service chief as its honorary vice-president; the discontinuation of that practice early in the 1950s was a turning point that is more obvious looking back than it was at the time.

WILDERNESS SETASIDES

There was no specific authority for the Forest Service to establish wilderness areas. Instead, as with most of what it did, the 1897 Organic Act allowed such setasides as part of its general delegation of powers to the secretary of Agriculture to regulate use. McArdle remembered that it was "late 1955 or early 1956" that Howard Zahnizer, executive director of the Wilderness Society, approached him about the need for wilderness legislation. The idea itself was not new, but the time had apparently come. McArdle and others in the Forest Service "tried to talk him out" of it. They were not "unalterably opposed" but felt that it was not "urgently needed" and the political timing was not good.

Zahnizer was not impressed by their arguments; his overwhelming concern was the apparent lack of permanence of a mere secretarial withdrawal. He "wouldn't buy" any of their arguments, insisting that legislation was essential. Later, Zahnizer and representatives from other groups that favored wilderness met in a Forest Service conference room to examine a draft that Zahnizer had brought with him. To McArdle's surprise, the draft received little support. "No one at this meeting was in favor of the draft legislation," and "Zahnie" left the meeting "dejected." But he went ahead, and it was Zahnizer's bill that Senator Humphrey introduced as S. 4013 on June 7, 1956. McArdle's interview includes a five-page Appendix E that lists the series of bills and revised versions that

Congress considered for eight years, as wilderness bumped along a rocky path toward "permanence."

Objections to wilderness legislation continued throughout the Eisenhower presidency, and Interior's opposition, particularly that of the National Park Service, was even stronger than that of the Department of Agriculture—that is, the Forest Service. According to McArdle, the Park Service did not want to consider any bill, but the Forest Service sought alternative language that would be acceptable. It took four years just to agree upon a definition of wilderness.

The Park Service's opposition was twofold. Publicly, the agency wanted Congress to wait until the Outdoor Recreation Resources Review Commission's report was completed, as "wilderness was only a minor part of the whole outdoor recreation picture and did not merit special legislation." Privately, Park Service Director Conrad Wirth told McArdle that "wilderness legislation would tie his hands in building roads and other developments needed in the national parks, and for this reason alone he would oppose legislation."

However, "the strongest, longest continued, and most effective opposition to wilderness legislation came from the forest industries." Other interests—mining, livestock, and highway construction—were also opposed. McArdle thought that the highway industry mounted "exceedingly effective opposition." Users of public lands such as sportsmen joined with the American Forestry Association to defeat wilderness legislation. The Society of American Foresters, by mail ballot, favored the continuation of wilderness creation by executive order, instead of congressional involvement.

When John Kennedy entered the White House, the political balance shifted. Both Agriculture Secretary Orville Freeman and Interior Secretary Stewart Udall "enthusiastically endorsed legislation to preserve wilderness." McArdle thought that the "big change was in Interior" because Agriculture had "been doing what we could to further the legislation." Udall's support was "very helpful," McArdle reported, but he added that "some of his people continued to undercut him. Don't ask me for any names, because I'm not going to give any."

As so often happens, personalities mattered. McArdle thought that Zahnizer "was about as thorough a gentleman as he ever knew." However, wilderness had become his "obsession." Senator Clinton P. Anderson of New Mexico, whose bill was adopted in 1964, reported to McArdle a recent visit by Zahnizer. "Anderson was really upset. He told me that if Zahnizer came up to beat on him just one more time, he was going to drop the whole effort." The senator said that he was already doing everything that he could do to advance the wilderness bill, but he was "fed up with constant yapping at him." Fortunately for wilderness preservation, Anderson was "never one to cherish a grudge," and McArdle "smoothed it over." McArdle believed that "without Clinton Anderson there would have been no wilderness law."

ANOTHER KIND OF MULTIPLE USE

McArdle lists among his "major accomplishments" the 1955 passage of the Multiple Use Mining Act. He credits his predecessor, Lyle Watts, for being the one who "really started action on mining claims." He filled 21 pages of single-spaced transcript to recount the story, making it the longest in the 252-page interview. His wilderness recollections came in a close second with 20, and the regulation issue required 17.

Mining on public lands predated creation of the national forests. During the 1890s, when Congress was figuring out just what the new forest reserves ought to be, it was clear that commodity interests were not to be adversely affected if they suddenly found themselves inside a federal forest. Therefore, the secretary of the Interior, and then Agriculture, was given authority to regulate occupancy and use of the forests; mining and other activities could continue and in fact were encouraged, but not to the detriment of the overall purposes of the forests themselves. The subsequent record shows a big gap between the high-minded notions of Congress and what happened on the ground.

To the Forest Service, the issue was not mining so much; rather, it was the staking of claims and what happened to the timber on the surface. The miner could extract valuable material from the ground but could legally use its trees only for mining purposes, such as pit props or even a cabin for the miner himself. He could work an unpatented claim but did not own the surface. When truckloads of high-value logs cut from an unpatented claim headed down the road for sale to a mill, therefore, Forest Service officers saw it as a type of theft.

The logging of unpatented claims had been going on as long as there had been a market for the logs. During the postwar years, two loosely related activities caused the number of claims on national forests to spike: the uranium boom to feed the nuclear industry brought prospectors out in droves, and "unscrupulous developers" staked claims along lakeshores and in similar areas where the market for summer homes was also booming. The endemic problem was now an epidemic; by 1955, new claims on national forests were being filed at a rate of five thousand per month. McArdle, as had Watts, believed that the 1872 Mining Law needed to be revisited.

McArdle had spent nearly all of his first six months as chief in the field, where he heard many reports about mining problems. The need for change was urgent, in part because of a "big upsurge" in mining claims. With Republicans in control of the White House and both houses of Congress, all of McArdle's congressional contacts advised him that the mining law would not be revised unless the mining industry itself could be brought into line. Representative Clifford Hope of Kansas, the new chairman of the House Agriculture Committee, introduced a trial-balloon bill in 1953 limited to claims on national forests, explaining to McArdle, "We probably won't get very far...but if we can stir up enough interest, it may throw a scare into the mining people." Nearly one hundred witnesses testified at Hope's hearing, mostly in support of his bill, and its opponents were "snowed under." Broad support had come from conservation groups, sporting clubs, professional organizations, and other members of Congress.

A high-profile scandal kept the issue alive in the media, and McArdle believed it added enough to existing pressures to prompt the mining industry to then support changing the mining law, with the American Mining Congress taking the lead. The Al Sarena Company was an Alabama firm that owned twenty-three heavily timbered, unpatented claims on the Rogue River National Forest in Oregon. The company applied to take the claims to patent so that it could cut the timber, a process that included the requirement to prove "adequate mineralization." Another part of the process required that the Bureau of Land Management ask the Forest Service for comment, and agency mineral examiners protested that fifteen of the twenty-three claims had no minerals. BLM accepted the Forest Service's findings, and Sarena's appeals to the secretary of the Interior were rejected. Then in January 1953, the new Eisenhower administration decided to allow the Sarena Company itself to provide new assays, which showed valuable minerals on the fifteen claims. Within a year, the company had its patents. It promptly sold two and one-half million board feet of timber, leaving, according to the company's estimate, another eighteen million. At no time was any mining undertaken.

This "sordid" tale of political maneuverings, the mysterious finding of assay samples, and "some pretty smoky undercover doings" kept Al Sarena on the nation's front page. There was "so much complaint to Congress" that appropriate House and Senate committees formed a joint committee and held hearings in Oregon and Washington, D.C. The mining industry was by now fully agreeable to certain revisions to the 1872 law but feared there might be "further and more radical changes." It limited its support to changes to "simplify enforcement and minimize bad-faith attempts" to extract minerals. McArdle reported that when President Eisenhower signed the bill into law on July 23, 1955, he called it "one of the most important conservation measures affecting public lands that has been enacted in many years."

INTERIOR AND RELATED AGENCIES

It seems rather strange that an agency in the Department of Agriculture would be included in the appropriations for the Department of the Interior and "Related Agencies." To many in the Forest Service, this is all part of its heritage dating from the 1905 transfer of authority over the national forests from Interior to Agriculture. After all, the General Land Office and later the Bureau of Land Management had retained jurisdiction over mineral rights and land surveys on the national forests, so it is easy to suppose that in Congress, appropriations measures had remained as they had been before 1905. But they did not; instead, it all happened a half-century later in 1955, during McArdle's watch.

McArdle first learned of the transfer of the Forest Service from the Agriculture appropriation to that of Interior from a "telephone call from Congressman Jamie Whitten," who was chairman of the appropriations subcommittee for the Department of Agriculture. Whitten himself had just learned of the transfer and that it had been ordered by the chairman of the full Appropriations Committee, Representative Clarence Cannon. Cannon explained to Whitten that he wanted all public works appropriations under a single subcommittee, which he would head, including transfer of the Bureau of Reclamation from Interior to Public Works. As things can go in Congress, Representative Michael Kirwan, who was chairman of the Interior appropriations subcommittee, then demanded "something" transferred to him that was about the same size as Reclamation, as measured in appropriation dollars. That "something" turned out to be the Forest Service. According to Whitten, it was the clerk of the committee who picked the Forest Service, whose budget was the closest in size to that of the Bureau of Reclamation.

The silliness of these switches was obvious, and there was backlash from other members of the House. McArdle recalled that Speaker Sam Rayburn, in an unusual gesture, "stepped down" from his chair to assure the House that the realignment of appropriations "was in no way to be construed as indicating a possible transfer of the Forest Service from Agriculture to the Department of the Interior."

McArdle then visited with Carl Hayden, chairman of the Senate Appropriations Committee. The chief asked the senator just "how this unprecedented business" would be handled in the Senate. To McArdle's dismay, Hayden said that he would "follow the lead" of the House, in accord with constitutional assignment of appropriations responsibility to that body. He added, "Don't worry about it, McArdle, because I'm going to be chairman of that subcommittee." In fact, in his opening statement during the Senate appropriations subcommittee hearing, Hayden stated for the record that his new subcommittee had been formed "for no other reason except to conform with a rearrangement of House Appropriations Committee responsibilities." During his interview years later, McArdle concluded his account of the "rearrangement" by stating, "I think it is good to put it in the record here because, of the Forest Service people, I think I'm the only one who knows what I've just told you."

Later on, Chief Robertson saw beneficial opportunities in being in Interior's budget. For example, he cited the time when the National Endowment for the Arts, one of the other "related agencies," had its budget cut by a Senate angry over support for supposedly pornographic art, which meant that more money was available within Interior's overall appropriation for the Forest Service. Chief Dombeck, too, saw advantages in being among several agencies in a relatively small department, rather than a small part of the much larger Agriculture budget.

ASSISTANT SECRETARY OF AGRICULTURE

While McArdle was chief, Secretary of Agriculture Ezra Taft Benson reorganized the department, among other things creating positions for assistant secretaries. In all previous administrations, the chief of the Forest Service had reported directly to the secretary. Now, the secretary would make decisions about the Forest Service based upon presentations by an assistant secretary. Thus, it was "very important that we have confidence in them" because the new system would not work well "if the assistant secretary was not competent."

McArdle provides no assessment of James Earl Coke, the first assistant secretary assigned authority over the Forest Service. However, he gives high marks to Ervin L. "Pete" Peterson, who succeeded Coke in 1954 and remained as assistant secretary until the end of Eisenhower's second term. He was a "lifesaver to the Forest Service in more ways than one, because of his interest in what were trying to do, his recognition of all the surrounding circumstances, and his willingness to make decisions and to back us up when he thought we were right." There was "nothing wishy-washy" about Peterson.

Before Peterson, no political appointee accompanied the chief when he testified on "important legislation." McArdle believed that he and "Pete" made a "good team," with Peterson providing him with political protection during hearings. This was "something new" to the Forest Service, and it "worked fine."

The chain of Forest Service command: Chief McArdle with Regional Forester J. Herbert Stone, Forest Supervisor Larry Barrett, and District Ranger Nevan McCollough of the Snoqualmie National Forest, probably late 1960. McCollough was one of the last "*Use Book* rangers," in that he was not a forestry school graduate. USDA Forest Service Photo.

DECENTRALIZATION

Gifford Pinchot decentralized the Forest Service in 1908 by creating inspection districts, known simply as districts (renamed regions in 1930). Since then, "local questions will be decided upon local grounds," as stated in the legendary *Use Book*, the concise forerunner to the modern-day, and voluminous, *Manual of Regulations*. Decentralization anecdotes abound, many no doubt apocryphal, such as Regional Forester Charlie Connaughton's supposed retort to Chief Cliff, "You do your job, and I'll do mine." Such stories aside, the agency is perhaps the most decentralized of all federal agencies. As McArdle observed, "Some agencies claim to be decentralized, but all of their decentralization is in Washington." Those agencies had their regional offices in Washington, and not in the field, as did the Forest Service.

Successful decentralization required three conditions. First, there "must be a thorough understanding of the organization's objectives, its goals, its policies, and programs." The second requirement was "effective delegation" of authority, or in today's vocabulary, no micromanagement. Third, routine inspections were necessary to ensure that tasks were being properly carried out and to provide feedback to those in Washington.

In his interview, McArdle recounted an early experience that impressed upon him the true sense of decentralization with its delegated authority. Even though a researcher at the time, he was named fire boss on a major fire. As he arrived with his crew, the fire broke out and he "saw a huge cloud of black smoke." He panicked and phoned the forest supervisor for advice. The tired response was, "I don't know, Mac. It's your fire." McArdle "learned then to accept delegated authority and responsibility."

There was a kind of centralization going on that McArdle did not like, a process that affected all agencies. Statutes that set eight-hour workdays as the standard were examples of change that was affecting Forest Service culture. Early in his career, McArdle recalled, his supervisor told him that he didn't have any time of his own, "that the government had it all." McArdle took that to heart and believed that when employees used all of their allotted vacation, "stagnation" sets in and there is a tendency "to forget the spirit of service." But he saw that overall, Forest Service people still had special qualities; "the clock has never ruled us in the Forest Service, and I don't think that it ever will."

NATIONAL FOREST DEVELOPMENT PLAN

"If we are to increase the use of the national forests for wood production, recreation, watershed, grazing, and wildlife, to meet increasing demands for all of these products and services, we're not going to do it without some effective planning." With that statement, McArdle introduced the National Forest Development Program for 1959. The Multiple Use–Sustained Yield Act was a year in the future, and McArdle was laying the groundwork for carrying it out by "getting activities in balance" and "giving equal consideration."

The agency took a "fresh look" at each program, given human population projections to the year 2000. The study also included short-range goals with a ten- to fifteen-year outlook. McArdle determined that "most of the imbalance in natural resource development can be attributed to policies and attitudes of budget people in the department and nationally. These budget people had no real understanding of need." Perhaps the development plan would alter such attitudes, and McArdle saw it as one of his "major accomplishments."

On September 21, 1961, President Kennedy sent to Congress a nineteen-page report, "A Development Program for the National Forests." The president stated that "this report modifies and supplements the 1959 National Forest Program submitted by the preceding administration." Major points were acceleration of timber harvest, adjustment of the road program, and substantial increases in recreation resources. As produced by the Department of Agriculture, the attached report stated, "the long-range timber goal for the National Forest System is an annual harvest on a sustained-yield basis of 21.1 billion board feet of saw timber by the year 2000," a doubling of the 1961 cut. The plan also proposed construction of 379,900 miles of new roads. Fifteen years later, Forest Service studies would reveal that the overcutting of old-growth timber had begun in the early 1960s under sustained-yield targets that were too high, but during McArdle's time, it all seemed not only possible but desirable. As it turned out, events closely related to changing societal values and new science would intervene, and the proposed, even higher, targets would not be even approximated.

McArdle retired on March 17, 1962, a year into the Kennedy administration. Edward P. Cliff, who succeeded him the next day, later recounted that "the manner in which [McArdle] succeeded in keeping the Forest Service in the hands of a professional chief [following Eisenhower's election], and then having a professional chief succeed him [during the early Kennedy administration], was one of the important contributions he made."

In retirement, McArdle was named executive director of the National Institute of Public Affairs and lectured at various colleges. He also served on the boards of several forestry organizations.

CHIEF EDWARD P. CLIFF

USDA FOREST SERVICE PHOTO

In their introduction, the interviewers of Edward P. Cliff comment on his "truly remarkable memory" and the absence of "any fear of the microphone." They added that the interview "expresses his sentiments in his own terms, and that he says exactly what he wants to say—no more and no less." Those who knew Cliff only at a distance remember a burly man who always had a curved-stem pipe in his mouth. But even those who never met him know that as chief, he encountered the first salvos of the environmental movement against his agency.

Cliff was born on September 3, 1909, in Heber City, Utah. At Utah State University's just-established forestry department in the School of Agriculture, one of his professors was Lyle F. Watts, who would in little more than a decade become Forest Service chief. Watts would remain an influence on Cliff during the early part of his forestry work. He mainly studied range subjects and was given a special research project in game management. Cliff graduated in 1931 and began his long Forest Service career as an assistant district ranger on the Wenatchee National Forest in Washington State.

In 1934, Cliff transferred to the regional office in Portland as one of the first wildlife specialists in the agency. His background in range management included

Source: Half a Century in Forest Conservation: A Biography and Oral History of Edward P. Cliff, *by Ronald B. Hartzer and David A. Clary. Washington, D.C.: Government Printing Office, 1982. Cliff was interviewed in 1980.*

an understanding of the sort of overpopulation problems that were a challenge for game management of the time. He also collaborated in writing the *Range Plant Handbook*. In 1939, he was moved to the Siskiyou National Forest as its supervisor and in 1942 became supervisor of the Fremont National Forest.

His next promotion and related move were to Washington, D.C., as assistant chief of the Division of Range Management, in 1944. Two years later he returned to Utah as assistant regional forester for the Division of Range and Wildlife Management. He then moved to Denver as assistant regional forester; he was promoted to assistant chief for National Forest Resource Management in 1952, the same year that McArdle had been appointed chief.

As assistant chief, Cliff was much involved with increased recreational use, mining issues, a steep increase in timber sales, and other aspects of the search for the proper balance of uses on the national forests. When the Multiple Use–Sustained Yield Act passed in 1960, he thought that it did not make any "substantial changes in the way we were trying to do business." He believed that the new law "reflected what we thought we should be doing."

SELECTION AS CHIEF

As had his predecessors, McArdle wanted to time his retirement to best ensure that the secretary of Agriculture would select his successor from inside the Forest Service. He had survived the significant shifts from Truman to Eisenhower, and then eight years later from Eisenhower to Kennedy. In early 1962, McArdle told Secretary of Agriculture Orville Freeman that he wished to retire. The chief then recommended three successors—three assistant chiefs—for the secretary's consideration: Edward C. Crafts; Arthur W. Greeley, whose father had been chief during the 1920s; and Cliff. The secretary asked each candidate to provide him with an essay "outlining what we thought were the principal problems and challenges facing the Forest Service"; a copy of Cliff's essay is appended to his interview. Freeman then directed an aide to examine credentials and contacted appropriate members of Congress.

President Kennedy was also involved in Cliff's selection. After a hearing before the Senate Appropriations Committee during McArdle's final days as chief, Carl Hayden, committee chairman, invited both McArdle and Cliff to his office for a chat. Hayden told them, "You know I called Jack [Kennedy] and told him I'd be very pleased if he would select someone from within the Forest Service to be chief." The president promised, "That will be done." Many years later, Freeman commented to Chief Dale Robertson that he had selected Cliff because he was

President John F. Kennedy dedicating the Pinchot Institute for Conservation Studies at Grey Towers, Milford, Pennsylvania, with Chief Cliff at his right and Secretary of Agriculture Orville Freeman at far right, 1963. Grey Towers remains an important Forest Service training and conference center. USDA Forest Service photo.

the "most chiefly" of the candidates. Cliff was appointed chief of the Forest Service on March 18, 1962.

Cliff believed that "any chief, to be successful in managing the Forest Service, has to be somewhat of an extension of the traditions and kind of leadership that preceded him." However, there was "no question" that a sitting chief could influence the agency's direction. He also saw that the chief was really "chief and staff" because most decisions were a consensus of views held by his "principal assistants." Thus, "no one man can really shape the destiny of the Forest Service." In contrast, McArdle recorded that "no matter now much advice I got before making a decision, when the time came to make a decision, I was the one who made it. It was not a vote by my staff people although I always paid particular attention to their views."

FORESTERS AND OTHERS

When Cliff was interviewed in 1980, he felt no need to touch on what is now called workforce diversity. Racial equality had become a legal requirement while

Deputy Chief for Research George Jemison

Agency Hiring Practices in the 1960s

The equal opportunity policy directed field officials to remedy what were considered to be limitations in their personnel management program. I was involved in that in a very critical way, and I made sure it involved our field research administrators.

For some reason, the secretary's office singled out Forest Service research as the unit that would be under special scrutiny to see that equal opportunity employment legislation and the employment of minorities were carried out effectively.

This scrutiny led to very critical reviews of our activities by the assistant secretary, even though statistics showed that we probably had the best record of any Forest Service unit in employing minority people. I made a careful personal survey of all our field units, and I found some rather deplorable situations. Not very many, but some. These were, of course, corrected. I also exerted a lot of time and effort in convincing our field administrators and our research station directors to do a better job in handling our equal employment opportunity programs.

We were instructed to go to every length to employ blacks, even to the point where I had to justify every promotion on the basis that we would try to fill in the opened position with a black. Of course, we couldn't find qualified people in very many cases.

Certain field station directors and I personally visited several black universities to make them aware of our program, and of the kind of people we wanted. We were hiring not just graduate foresters, but also physicists, sociologists, and in thirty-five other research disciplines.

There were qualified people around, if you could find them. I can't say that those university visits led us directly to highly competent blacks whom we hired. We even had some rather unnerving experiences where we got taken by blacks we were - trying to hire. And I'm not proud of this, but we may have hired a few who were really not qualified for the job, just because of the pressure we were under. But the program and all this pressure, I would have to say, did some good overall. I think it helped the organization, made us more conscious of what we should have been doing, and corrected some bad situations.

We placed a full-time forester, paid out of research funds, at Tuskegee Institute to counsel students. His job was to tell them what forestry was all about, identify an aptitude or an interest in forestry or biologically oriented programs, give them summer jobs, and manage the six-thousand-acre forest property at Tuskegee. That program, I understand, is still going on. Again, I can't be sure that it produced many black forestry enrollees, but I'm sure that it was worthwhile.

From An Interview with George M. Jemison, *by Elwood R. Maunder. Durham, North Carolina: Forest History Society, 1977.*

he served as chief, and at the very end of his administration, the National Environmental Policy Act had spawned interdisciplinary involvement in the technical aspects of Forest Service programs. However, his comments focused on the increasing number of disciplines; there is no mention of women and minorities in his interview in 1980.

"It's been charged that the Forest Service has been ingrown and that we're primarily interested in professional foresters," that "unless you were a professional forester, there was no chance for advancement in the Forest Service." Cliff added, "Of course that is not true." He believed that the accusation could be "refuted" by looking at the record, the "history of many of the people who now hold top positions."

Cliff's list began with R. Max Peterson, who had been appointed chief in 1979. Peterson was an engineer with training in administrative management, and he had "long experience" in the Forest Service. Hamilton K. Pyles had risen to the position of deputy chief, even though his training had been in anthropology. Like Peterson, Pyles "had a Forest Service background." Regional Forester George James was "basically" an engineer "but with intimate knowledge" who had "come up through the ranks." Finally, there was Jay Price, "one of our great regional foresters," who also was an engineer. Cliff ended his summary, "I could go on and on." However, he did not provide additional examples, and from today's vantage point it would seem that only those who had demonstrated a commitment to traditional Forest Service culture were selected for advancement, a point well made at the time Cliff was chief by Herbert Kaufman in his seminal *The Forest Ranger*.

SECRETARY OF AGRICULTURE

McArdle had arranged for Cliff to accompany Secretary Freeman on a field trip while he was still an assistant chief. Their tour of the Grand Mesa National Forest was "the beginning of a long and close friendship." The Grand Mesa was a good place to explain to the secretary just what multiple use was, except the "weak link" was its very modest timber program. However, it was the site of "some of our bitterest contests over livestock grazing," and Freeman could see for himself just what range and watershed rehabilitation looked like on the ground. In the evenings, they "did a little fishing." After he became chief, Cliff and Freeman made a "number of trips" in wilderness areas and areas outside the national forests where the agency had cooperative rural development programs.

Cliff remembered Freeman as a skilled administrator. He had adopted the practice of holding informal meetings every working morning that included

Secretary of Agriculture Earl Butz with Cliff during a visit to Forest Service headquarters, 1972. USDA Forest Service photo.

assistant secretaries, principal bureau chiefs, and senior staff. They met in an executive dining room before the "usual opening hour of the office." Everyone felt free to discuss "particular problems that they might have, to ask questions, to get answers, and to keep up-to-date with what was going on in the department generally." The secretary participated fully in the give-and-take, and it created "a family atmosphere within the department that I have not seen equaled before or since." Cliff also had "very strong support" from Assistant Secretary John Baker, "who demanded much of the agency heads, and we all tried to respond."

Some personnel appointments that chiefs make require secretarial approval, and Cliff was successful in gaining Freeman's support for "all of the key appointments and promotions that I made." Subsequent secretaries—Clifford Hardin and Earl Butz—also approved Cliff's personnel decisions "without exception." When he retired in 1972, all deputy chiefs, experiment station directors, and division directors in the Washington office had been selected by Cliff, leaving the agency in "strong hands." He thought it "one of my more notable accomplishments."

Cliff had the "full responsibility for carrying the full leadership of the Forest Service" through the transition from the Kennedy and Johnson administrations to that of Richard Nixon. As had McArdle, Cliff tried to pick a retirement date that would provide the least opportunity for his replacement to be selected from outside the Forest Service. Cliff knew "for a fact because Freeman told me" that the outgoing secretary had recommended to incoming Secretary Clifford Hardin that he be retained. Cliff did not know whether serious thought had been given to removing him, but he worked his final three years during Nixon's presidency.

Assistant Secretary Tom Cowden became Cliff's primary route to Secretary Hardin. Cowden was responsible for the Forest Service, Soil Conservation Service, and the Rural Electrification Administration. He held frequent meetings with his bureau chiefs, and he and Cliff developed a "close and cordial working relationship." Cowden's efforts produced strong secretarial support for Cliff and the others. Cliff had "no complaints." As an example, he recalled that when timber industry officials had asked for a meeting with Hardin to obtain "modification" of some Forest Service policies and practices, the secretary insisted that Cliff be present to "express his viewpoints." There were several of "these confrontations," and Cliff received "complete and strong backing." Cowden remained as assistant secretary when Earl Butz replaced Hardin. Cliff retired early in Butz's tenure, but the "degree of teamwork and cooperation and mutual trust was very strong."

DEPARTMENT OF THE INTERIOR

In his essay on what he would do as chief, prepared at Secretary Freeman's request, Cliff wrote that "the Forest Service needed to get off the defensive in its dealings with the Department of the Interior." Mainly he was thinking about the historical fact that many national parks had been created from national forests, initiatives that "almost invariably" had come from Interior. Facing whomever Freeman selected would be proposals for the Sawtooth and North Cascades national parks.

As the record shows, the Forest Service "resisted many of these changes" but not all of them. Cliff remembered that the Forest Service and the Park Service had been able to agree upon boundary shifts that worked to the advantage of both agencies. He used Yellowstone Park as a model of working together. He also remembered an agreement for a land exchange between Sequoia National Park and Sequoia National Forest. The Sierra Club "intervened," however, and even though the national forest area was transferred to the park, the Sierra Club was able to block the reciprocal swap from park to forest.

To Cliff, new programs such as the Visitor Information Service were good examples of being less defensive. Now the Forest Service could offer "a better and larger variety of visitor services to recreationists." Plus, the Park Service trained Forest Service staffers to handle this new venture. With talent in hand, the Forest Service proposed that instead of a Sawtooth National Park, a national recreation area administered by the Forest Service would meet the same need. This "put to rest the agitation for a national park, for a time at least."

High-elevation portions of the national forests in the North Cascades had long been classified as a "limited area," part of a vast reservoir of potential wilderness areas. Although some land was ultimately classified as wilderness, advocacy for a national park grew. Cliff accompanied Secretary Freeman, Interior Secretary Stuart Udall, Park Service Director George Hartzog, and Washington Senator Henry M. "Scoop" Jackson on a helicopter survey of the proposed park area, followed by a horseback trip with Freeman and his wife. Cliff admits that "we were on the defensive. We were trying to defend our turf, and didn't defend it successfully."

The debate over a new North Cascades National Park versus an existing wilderness area was bucked to the White House by Freeman, who had "stood shoulder to shoulder with the Forest Service in resisting the transfer. President Johnson listened to Freeman's arguments and then said, 'Well, I think we owe one to Scoop, so I'm going to go along with Interior.'" Thus, the paying of a political debt to a senior senator (who later would play a major role in advancing the National Environmental Policy Act) turned the tide, and once again, the carefully kept Forest Service scorecard showed that it had "lost" to the National Park Service.

WILDERNESS ACT OF 1964

By the time the much-amended wilderness bill finally passed in 1964, Freeman and Cliff had both testified in support. To Cliff, it was ironic that after passage, "it became more complicated, more difficult to create wilderness areas. It really did." Before the act, the chief "by the stroke of a pen" could create wild areas between 5,000 and 100,000 acres, and the secretary had similar authority to establish wilderness areas larger than 100,000 acres. By the time of the Wilderness Act, "the Forest Service had established a system comprising a total area of 14.6 million acres of wilderness and primitive areas."

The Wilderness Act included for the first time land in national parks and areas managed by the Bureau of Land Management. But as Cliff noted, "times were changing." It had been so difficult to gain consensus on earlier administrative wilderness setasides that by 1964, such decisions "should properly be relegated to Congress and subject to all of the review and the public hearings and the study that has to be made before a classification of that kind is made."

Cliff "made the decision that we would move forward and study roadless areas," by a process called Roadless Area Review and Evaluation, or RARE. There were still many areas within the national forests that had yet to be logged and had potential for wilderness classification. "The Sierra Club, the Wilderness Society, and other advocates of wilderness were challenging nearly every move the Forest Service was making towards any development," such as road building, timber harvest, and other disruptive uses, "if the lands were presently roadless."

Wilderness advocates obtained a major advantage "as a result of a number of court cases" that overturned the long-held "legal doctrine of sovereign immunity," which shielded the executive branch from most lawsuits. With the doctrine overturned, administrative decisions "then became subject to challenge in the courts. This created a whole new ball game." Legal challenge was added to the "other avenues of delaying actions." In 1969, when Senator Jackson was chairman of the Committee for Interior and Insular Affairs, he saw the need for including the requirement for an environmental impact statement in the language under consideration for the National Environmental Policy Act. Cliff soon learned that these "impact statements could be and often were challenged and questioned, causing further delays. It fell to John McGuire, Cliff's successor, to carry out the RARE studies.

Cliff concluded his wilderness recollections: "It really bothers me that the Forest Service doesn't seem to get any credit...for the many, many things that we have done in establishing and fighting for wilderness....Enough said."

ENVIRONMENTAL MOVEMENT

Edward Crafts, who left his post as assistant chief to become first director of the new Bureau of Outdoor Recreation, later stated that "the Forest Service was slow and insensitive to the wakening national awareness of our environment." Cliff remarked, "I don't disagree entirely with that," adding, "I can admit, in my own reactions, I didn't fully or quickly recognize the total potential impact of the strength of the opposition that was developing against such things as clearcutting." He went on, "Clearcutting is a perfectly sound silvicultural management system if it's properly applied."

Properly applied or not, clearcutting had become anathema to more and more people. Cliff recalled that early in the Nixon administration, he "had a visitation from two or three people from the Monongahela National Forest area, who came in to protest clearcutting." Hunters of wild turkey, they resented that their favorite areas had been cut and "demanded that this be stopped." Although Cliff sympathized with their point of view, his "initial reaction was, What's more important, the personal pleasure of a small handful of people for turkey hunting or the utilization of this resource for production of jobs and raw material?" Cliff admitted not recognizing the strength of the opposition and "didn't pay as much attention to that protest as it deserved."

"It soon got beyond the turkey hunting issue." The "militants," as Cliff characterized them, engaged the Izaak Walton League in the clearcutting conflict. Shortly, *Izaak Walton League v. Butz* was decided for the plaintiffs and "challenged the whole legal basis for timber management on the national forests." If he had it to do all over again, Cliff doubted he would size up the Monongahela issue as more than a "very self-centered protest from a very small segment of the population. . . . What I didn't realize is how potent they could be in expanding this protest." He doubted that anyone had really understood the astonishing pace of environmental events. Despite sincere attempts to provide proper management of the national forests, "many people have not fully adopted or subscribed to the principle of multiple use. Some even object to any timber harvest or any development on any national forest land that's not now developed."

The high-impact Monongahela decision came down a decade after an even higher-impact event, the 1962 publication of Rachel Carson's *Silent Spring*. The year was also Cliff's first as chief. Cliff thought that Carson had "performed a great public service . . . by calling the attention of the public to the potential hazards of promiscuous use of pesticides or other chemicals in our environment." Carson was a "talented biologist, a good writer, and she had a message to give." Cliff had "no quarrel" with Carson, but he was much bothered by the others—"people who get emotional"—who went "way beyond" her and urged "total elimination" of pesticides.

Carson was invited to a White House ceremony, and over the strong objections of the Forest Service and other agencies, the president's Science Advisory Committee in 1963 recommended much more stringent safeguards for pesticide use. DDT with its high effectiveness but long persistency received most of the attention; the Forest Service insisted that the agent needed to be used until alternatives could be found. Toward that end, the Forest Service research arm, supported by the largest appropriation for a specific study to that time, very vigorously expanded its efforts to develop chemical replacements for DDT, as well as developing biological controls. In 1973, the Environmental Protection Agency yielded to unprecedented political pressure and very reluctantly issued the Forest

Service a permit to use DDT to control a tussock moth infestation in eastern Oregon. This was the last time that DDT was legally used in the United States. By that time, EPA had approved Zectran as the pesticide to replace DDT, after screening 130 chemicals over six years. In 1975, by mail ballot, members of the American Forestry Association listed Carson among the "top ten most important conservationists in American history."

CLEARCUTTING

Cliff's thoughts and recollections about clearcutting above were expressed mainly within the context of multiple use and the beginning of the environmental movement. After discussing several other topics, he returned to clearcutting, and his comments fill an additional thirteen single-spaced pages. Arithmetical analyses are not the best way to measure historical significance, but it is fair to say that Cliff and the Forest Service during his time as chief saw clearcutting as a central issue. Despite the stout defense that clearcutting was an "acceptable silvicultural practice"—language inserted in the National Forest Management Act of 1976—more and more of the public liked it less and less. The "clearcutting problem" just would not go away.

Except within forestry circles, clearcutting has never been popular. In fact, the 1897 Organic Act included language that trees were to be "marked and designated," a method used for selective logging but not clearcutting. The Monongahela judge had pointed to this exact phrase when deciding that clearcutting was in violation of the early law. When Bernhard E. Fernow, Gifford Pinchot's predecessor at the U.S. Division of Forestry, as dean of forestry at Cornell had supervised clearcuts in New York State, the legislature zeroed his appropriation and canceled the program. Minutes of Forest Service chief and staff meetings during the 1920s show that some agency leaders were opposed to hiding clearcuts from view by leaving screens of standing trees along roads; cutting to the road and putting up explanatory signs was the more honest way to go. Even though silviculture texts that described clearcutting as an acceptable practice continued to pile up, an even larger pile of books and articles—some of them "emotional"—that opposed clearcutting cast the bigger shadow. In the early 1990s, Chief Dale Robertson would announce that clearcutting was no longer a "standard practice" on the national forests.

But it definitely was a standard practice when Cliff was chief, mainly used to log shade-intolerant species and timber in rugged terrain where road building was both difficult and damaging. The euphemisms "even-aged management" and "monoculture" quickly were adopted by clearcutting opponents as pejoratives. Cliff recounted that the Forest Service implemented internal audits of not only the Monongahela situation but also clearcutting on the Bitterroot National Forest. The Bitterroot controversy received much commentary and analysis in forestry literature, especially in *American Forests*, published by the American Forestry Association. The audit teams included scientists, economists, silviculturists, and recreation and wildlife specialists. For the Bitterroot especially, the teams concluded "that the quality of management was being sacrificed to obtain quantity of production, and that the program was out of balance." Cliff then announced that "if this were the case," then the Forest Service would "no longer sacrifice quality for quantity of production, even if we had to reduce the objectives." The western press and forest industry outlets criticized his decision "quite sharply."

The timber industry, "still fearful of the specter of forest regulation," gave Congress little incentive to increase the appropriation for nontimber programs. Even though Cliff admitted that Forest Service programs had long been "out of balance" in favoring the commodity production side, Congress was to blame. Studies clearly support his contention that although Congress routinely provided full funding for timber management, programs in recreation, watershed, and wildlife would receive less than half the requested amount. The problem was twofold: just what mix of uses at what levels would achieve "balance," and then how to get the White House and then Congress to go along. The debate continues.

To get an "outside look by reputable scientists and foresters at clearcutting practices," Cliff formed the "Deans Committee." The committee of forestry school deans, including Arnold Bolle from the University of Montana, who had been very critical of the Bitterroot operations, "largely supported the principle of even-aged management where it was most appropriate. In other words, it was supportive of the general thrust of Forest Service practices." This sort of "unbiased outside look" was a "valuable service."

Senator Frank Church of Idaho held hearings on the Bitterroot and similar programs in Wyoming. Cliff of course testified and "acknowledged some of the mistakes that had been made." He also pointed out "some of the successes and strong points." The committee then issued guidelines on clearcutting, which the Forest Service adopted. Cliff thought that was an important turning point, as the "crescendo" over clearcutting died down. Later, when Congress was drafting the National Forest Management Act, the Church guidelines were pretty much incorporated.

REORGANIZATION

There is no mention of the 1966 Forest Service reorganization in the interview text; however, two related documents were placed in an appendix. At the time of the interview, the agency was unable to locate a copy of the 1966 organizational chart, but it provided one for 1974 with assurances that in all important aspects, it was identical to the earlier version.

The first document is a memorandum from Max P. Reid of the USDA Office of Personnel to Deputy Chief Clare Hendee, dated November 22, 1965. Reid stated, "The revised organizational chart for Forest Service was concurred in by Assistant Secretary Robertson and approved by Assistant Secretary Baker...This chart has been returned to the Forest Service." The major points were establishment of the position of associate chief, vacant since 1944; creation of northeastern and southeastern area offices for State and Private Forestry; abolishment of Region 7 and transfer of its programs to Regions 8 and 9; and abolishment of the Central States Experiment Station and transfer of its programs to the Northeast and Lake States stations.

The second document is from Chief Cliff to regional foresters, area directors, and Washington office staff, dated March 14, 1966. Cliff began, "The Washington Office has made significant organization changes during the past several months. These changes were prompted by a need to keep our organization structure abreast with present day program needs." Too, there was need to "sharpen the line and staff relationships" and clarify authorization for signing documents for the agency.

Changes and additions to the earlier approved organizational chart include merging the positions of assistant chief for National Forest Resource Management and assistant chief for National Forest Protection and Development into deputy chief for National Forest System. The deputy chief was to be assisted by two associate deputies. Later reorganizations would establish associate deputies for all deputy chiefs.

Signing authority was detailed; the chief and associate chief would sign for "issues involving the entire Forest Service. Deputy chiefs, however, had line authority only "within area of assigned responsibility," and so on down the line to associate deputies and division directors. Branch chiefs had no signing authority except as "acting," which would be a temporary authorization to sign in their superiors' absence. Cliff added, "Field units will establish operating procedures similar to those outlined above."

This sort of nuts-and-bolts effort was essential to a highly decentralized organization. If authorities were not clear, the timid would avoid or delay necessary decisions while the more confident—perhaps even aggressive—might go beyond their authority.

Cliff retired in 1972 but remained active as a forestry and land-use consultant. After a stint with the National Materials Policy Commission, he accepted a series of international assignments from United Nations organizations that included travel to twenty-one countries.

CHIEF JOHN R. McGUIRE

USDA FOREST SERVICE PHOTO

John R. McGuire was born in Milwaukee on April 20, 1916. He earned a B.S. in forestry from the University of Minnesota in 1939. During his undergraduate years, he held a variety of temporary assignments with the Forest Service, and he earned his master's degree in forestry from Yale University in 1941 while working at a Forest Service research facility on campus. Following the outbreak of World War II, he served four years in the Army in the Pacific theater, rising to the rank of captain in command of the Eighth Engineering Battalion. After mustering out in 1946, he returned to the Forest Service at its Northeastern Forest Experiment Station in Orono, Maine.

In 1950, McGuire was transferred to Upper Darby, Pennsylvania, as a mid-ranked forest economist. In 1955, he spent nine months in the Washington office, the first of many such assignments as he developed an ever-broader horizon. His stint at Upper Darby allowed him to pursue additional graduate studies in economics at the University of Pennsylvania, where he completed the required course work for a Ph.D. but not the dissertation.

His career broadened again in 1957 with assignment to the Division of Forest Economics at the Pacific Southwest Forest and Range Experiment Station at

Source: An Interview with John R. McGuire: Forest Service Chief, 1972–1979, *by Harold K. Steen. Durham, North Carolina: Forest History Society, revised 2004. McGuire was interviewed in 1988.*

Berkeley. The following year he was named chief of the division. He then moved to Washington, D.C., in 1962 as part of a promotion to assistant to the assistant chief for research; a little more than a year later he returned to Berkeley as director of the Pacific Southwest station.

After serving as station director for four years, in 1967 McGuire returned to Washington as deputy chief for Programs and Legislation, a job that involved him in issues far beyond the esoteric world of research. Too, as deputy chief he became well acquainted with members of Congress, developing relationships that later played a material part in his selection as chief.

McGuire was reserved and could wield authority when necessary. Popular with the press, he was referred to as the Jimmy Stewart of the forests, after the famous actor whose trademark was slow-speaking honesty. Chief Peterson described McGuire as "fairly quiet and nonconfrontational, so was an ideal person to have as a front man." Chief Dombeck referred to him as "a man of few words, but when he speaks, you'd better listen." To a question inspired by Herbert Kaufman's *The Administrative Behavior of Federal Bureau Chiefs*, when McGuire was asked how firmly a chief can exercise authority, he responded, "You can do something pretty drastic if you want to." However, in most situations it was preferable only to comment on a situation rather than to give orders; those within hearing distance, McGuire believed, would be responsive to the comment and take corrective action.

Although loyal to the Forest Service and its traditions, McGuire placed public service at an even higher level. "My attitude, and perhaps it was different than some, was not that the professional simply knows best, but rather it was more an attitude that since these national forests are public property, the public can decide to manage them any way at all." It was important to find out what the public wants done "and then do it."

SELECTION AS CHIEF

As he approached the minimum retirement age of fifty-five, McGuire began giving serious consideration to leaving the agency. Unexpectedly, however, he was promoted to associate chief in June 1971, and for the first time he realized that he was in direct line to be appointed chief when Cliff retired. The following April 28, McGuire became the tenth chief of the Forest Service.

In his interview, Cliff stated, "I only recommended one person for the position of chief. I had a high regard for John McGuire as a man and as a professional, or I wouldn't have made the recommendation I did." However, Cliff had never talked to his associate chief about it, so McGuire was not familiar with the specifics of

KEEP YOUR EYES OVER THERE, JOHN!

Cartoonist Rudolph Wendelin portraying the multiple challenges that Chief McGuire faced in 1972 as he assumed the chief's role. Former Chief Cliff, pipe firmly in place, points the way. McGuire is shown walking a high rope trying to balance the needs of the station directors and regional foresters while wearing a life preserver entitled "STAFF." Lurking below in the turbulent waters are the various issues represented by sharks, including pollution; pesticides; timber, grazing, mineral, and recreation demands; clearcutting; wilderness conflicts; funding; manpower ceilings; and reorganization. The illustration was signed by forty-three colleagues. Courtesy of Marjory McGuire.

just how he was selected as chief. McGuire knew Secretary of Agriculture Earl Butz from their joint appearances before congressional committees, and he knew many departmental staff. "Butz thought I ought to pass muster with the White House, so I went over and talked to the man who handled personnel for President Nixon." There were other White House staffers present during the interview, whom he also knew from previous dealings. "They never did ask me if I believed in the tenets of Republicanism or anything ideological." Mainly, the White House was concerned that there not be "any objection" from the Hill—that is, objections from Republicans on the Interior and Agriculture committees. Since McGuire was well known to key senior Republicans, and apparently highly regarded as well, his appointment went ahead.

The process taught him an important lesson: a chief must be able to work with the people that his predecessor leaves him. The deputy chiefs and regional foresters already in place form a talent pool that become the source of most of his selections for promotion. Perhaps too modestly, McGuire believed that he was selected primarily because there was no one else of the proper age and with the requisite experience.

During this period, Senator Henry M. Jackson had drafted a bill to place the Forest Service chief and National Park Service and Bureau of Land Management directors in the category of presidential appointees, to give the Senate official influence under its constitutional advice-and-consent authorities. Jackson was upset by recent appointments of "political people" to head Interior agencies, such as Nixon's campaign advanceman Ronald H. Walker to be Park Service director. McGuire felt that such a law would make the "job quite political" not only because of senatorial influence but also because it would then be easier for each incoming president to replace the Forest Service chief.

McGuire had learned that Saturday mornings were usually a "good time" to visit the Hill, as senators and representatives were alone in their offices and appointments were not necessary. He dropped in on Senator Jackson to persuade him to withdraw his bill, to "leave things alone for the time being" because it would be better to leave the chief's position under Civil Service unless the political situation "got a lot worse." Jackson agreed, and later on the Carter administration sponsored the Civil Service Reform Act, creating the Senior Executive Service. McGuire observed in 1988 that SES had kept the chief's job under Civil Service and, compared with most agency heads, "free of political favoritism."

EXECUTIVE BRANCH RELATIONSHIPS

The president and the advisers he selects are a significant influence at the bureau level. A new administration means a new secretary of Agriculture and a new assistant secretary who is the chief's direct political boss. Too, there is a shift in attitude in the Office of Management and Budget and other federal agencies. If the president is of a different party than the congressional majority, congressional staff tend to bypass political appointees and deal directly with the agency. Also, each president has his own management style.

During the Nixon administration, there were "a few" people in the White House who were "particularly interested in Forest Service matters." They tended to bypass the department and talk directly to the agency, resulting in an unusual degree of contact between the White House and the Forest Service. This direct contact extended to White House agencies such as OMB, the Council on Environmental Quality, and the Council of Economic Advisers.

McGuire got along well with Secretary Butz, who was easy to talk to, had ideas about the Forest Service, and tried to build a social relationship among his own staff but also with agency heads. The chief even accompanied the secretary to nonforestry meetings, such as a visit to the *New York Times* and trade negotiations with the Japanese. It was also to the good of the Forest Service that the secretary had a strong relationship with President Nixon; some cabinet members would only rarely see the president alone.

Assistant secretaries may fill a staff or a line function or both, but Butz used them mainly as staff. Assistant Secretary Tom Cowden, who was on paper McGuire's direct political boss, "wasn't all that interested in Forest Service matters and mainly wanted us to keep out of trouble." As a result, the agency dealt directly with the White House during those years.

Jimmy Carter had campaigned against the bureaucracy, causing "great distrust" at the White House staff level. This situation ended Forest Service entrée to the White House, as the chain of command reverted to political channels. Whereas he had had comfortable conversations with President Jerry Ford when the chief accompanied Secretary Butz to the executive mansion, McGuire never had an opportunity to sit down with Carter. His only meeting with Carter involved the traditional presidential tree planting at the White House. Carter had contacted the state forester of Georgia to provide the tree, an act that suggested to McGuire that the president was not really aware of the Forest Service. The state forester in turn contacted the chief, and the two of them arranged for the planting.

Woodsy Owl, Secretary of Agriculture Earl Butz, and Chief McGuire, 1973. Woodsy is the Forest Service symbol for environmental awareness. USDA Forest Service photo.

Without White House access, it was very difficult to talk to people like Stuart Eizenstat, who was Carter's chief of staff for domestic affairs. Eizenstat's assistants tended to be environmentalists, "who had pretty much warned him against the Forest Service." Once, "out of the blue," he accused the Forest Service of undercutting the administration because that is what he had been told. McGuire's only recourse was to assure White House staffers that his agency was not political and was not opposed to the Carter administration.

McGuire was never sure just where he stood with Carter's secretary of Agriculture, Robert Bergland. Assistant Secretary Rupert Cutler was the "environmentalist candidate appointed to exercise line authority over the Forest Service and the Soil Conservation Service," more than to provide a staff function. Since the relationship between the secretary and Cutler was unclear, McGuire "always felt free to go directly to Bergland about any problem."

The assistant secretary "used to meet with the environmentalists every week or so. He had an assistant whose principal job seemed to be to keep in touch with them." Cutler "was always coming up with press clippings or something that some environmentalists had sent him from the West that were critical of the Forest Service." However, the upside was that as Cutler's interest lay mainly with wilderness and other environmental issues, there was a large array of topics McGuire could discuss with the secretary without making the assistant secretary feel he was being bypassed. McGuire suspected that Cutler, who as a political appointee lacked tenure, wanted to be chief. However, when the McGuire interview was published in excerpt form and Cutler saw that statement, he strongly denied that he had ever aimed to become Forest Service chief.

WORKING WITH CONGRESS

McGuire often found it necessary to testify before congressional committees on issues—below-cost timber sales, for example—that had highly technical aspects. For that, the chief "might get fifteen minutes." A committee member might read a series of questions prepared by his staff, ending each, "Is that not correct?" If McGuire could answer "yes" to each, then the technical material would be in the record.

McGuire's testimony was supported by a staff-prepared book of information, up to an inch or two in thickness. Preparation consisted mainly of gaining familiarity with the book so that he could quickly turn to the correct pages when responding to a question. The testimony would be accompanied by a report from the administration that had been cleared by OMB and other interested agencies. The report contained the administration's position on the issue and assured the committee that the administration was "behind the witness." Since the report was ordinarily sent in advance of the testimony, the witness usually offered to summarize his statement and then be available for questions. At times an assistant secretary would lead off, and the chief's role would be to handle technical questions.

Political scientist Herbert Kaufman had examined the Forest Service and reported that during the study period, the agency had testified to twenty-three different congressional committees; the chief testified forty percent of the time, and political appointees or the associate and deputy chiefs provided the balance of testimony. As McGuire pointed out, "There's a lot of legislation that is not major policy, and there's no objection on the part of the committee if someone acting for the chief goes up and testifies."

Testimony provided by political appointees was usually drafted by the Forest Service and cleared with OMB and other agencies, producing a joint product. Thus, the testimony rarely included surprises, but "the ticklish part comes when the questioning starts." For example, if the questioner is of a different party than the assistant secretary, there might be some "political hay" made. At times the questioner was antagonistic toward the appointee and ignored him, directing questions directly to the chief instead. That approach "can get a little embarrassing."

Everyone well understood that the chief might not agree with the official position of the administration on an issue, such as certain budget line items, but that he could not openly object. McGuire recalled that experienced members of Congress would try to "draw [me] out" by asking for a personal opinion, and then "the witness has to respond." The administration understood that he "can't lie," and he would not get into trouble. This approach was most often used during budget sessions, with the result that the Forest Service frequently received an appropriation larger than what had been in the president's budget. But McGuire's favorite example was his testimony on the National Forest Management Act. Congressman James Weaver asked, "What, if you were writing this bill, in your personal opinion, would you put into it?"

With that opening, McGuire listed a "half a dozen things" that he would like to see added, "things that wouldn't have come up otherwise." The most important addition was making the national forests statutory. Unlike the national parks, which had been created by statute, most national forests had been established by presidential proclamation—and could be "disestablished" by proclamation just as easily. Thanks to McGuire's "personal opinion," under NFMA, a "president cannot move a national forest out of the National Forest System and back into the unreserved public domain."

Formal committee proceedings ordinarily featured deferential witnesses before powerful members of Congress. If in private the chief had developed a cordial relationship with an individual member, however, discussions could proceed as among equals. For example, "you can call a member by his first name, and he'll call you by your first."

OTHER AGENCIES

The Office of Management and Budget, created from the Bureau of the Budget by executive order during the Nixon administration, was "beginning to apply more economic analysis to budget requests." Too, OMB employed more people "competent to ask the right questions," which included recruiting staff from the agencies themselves. The Forest Service encouraged this approach and went beyond to encourage the Council on Environmental Quality and others to do likewise. During the Lyndon Johnson administration, the Bureau of the Budget gave more attention to budget details at the agency level; "earlier it was more a matter of setting overall targets for the department, and then letting the department figure out how to spend the money." Hence the increasing importance of OMB staffers who could "ask the right questions."

The desire for accountability also involved the General Accounting Office, the congressional agency that investigates activities of the executive branch in response to a request from a member of Congress. In a "complicated case," GAO would "set up shop" within the Forest Service. The agency would provide office space, and the GAO person would be given full access to files to "dig out whatever he wants." Ordinarily, GAO provided draft reports to the agency for comment, which would then be included in the published version.

In general, McGuire felt that "GAO has fairly competent people, but they are stretched pretty thin sometimes, and they do get into areas where they don't have a great deal of expertise." For the agency, "You've got to start with GAO people who may know nothing about the subject and sort of educate them and bring them along."

Relations with the other land management agencies in the Department of the Interior vary "with time and with the agency and with the personalities involved." The "perception of interagency conflict" bothered McGuire, so he tried to arrange a "regular schedule of meetings" with his counterparts in the Bureau of Land Management, the National Park Service, and the Fish and Wildlife Service. They would have lunch every few months, and their staffs would draft the agendas and in other ways also get to know each other. McGuire was "lucky with Fish and Wildlife and BLM, despite the continuing turnover in leadership, and we kept the meetings going." However, with the Park Service, "we never did quite succeed in reaching the same kind of accommodation."

THE CHIEF'S DAY

As chief, McGuire spent half his time in "intelligence gathering"—attending meetings, inspecting field operations, and reading staff summaries of issues. Nearly a third of his day was spent with "external relations," such as giving speeches to citizen and professional groups and working with Congress and the executive branch. The balance of his time was devoted to "motivating the workforce."

Each day began with a brief chief and staff meeting, a "show and tell" where "everybody would say what's coming up today." Cliff had held more formal weekly meetings that were useful for decision making but less so for "exchanging information and for staff coordination." Too, the meetings informed McGuire of what was "currently urgent" and provided clues to where he ought to focus his reading efforts for that day. Minutes were distributed "to everybody in the upper echelons," including the field. Often regional foresters and station directors would phone in with their thoughts about a particular topic. McGuire noted that from an "intelligence standpoint," the daily sessions worked pretty well. The potential downside was a tendency to put directives in the minutes, which could lead to a lack of coordination with more formal orders. Directives belonged in the Forest Service *Manual of Regulations*.

McGuire quickly discovered that as chief, he had to be "quite selective" in the material he read. For example, while he was deputy chief for legislative affairs, he tried to read the full *Congressional Record* each day. With responsibility for the whole of Forest Service activities, he now had to rely on staff recommendations about the *Record* as well as his other reading, such as internal staff reports, which in themselves could be "quite voluminous." Too, he had a lot of speech drafts to review; he subdued his impulse to heavily edit the text and tended more or less to use what his speechwriters provided.

Chiefs travel an astonishing amount to maintain familiarity with the complex issues and activities on the vast National Forest System. McGuire sought ways to moderate his travel, even though there was "somebody who's acting in his place back in Washington and who signs the important mail and keeps in touch" by telephone. However, the typical acting chief doesn't feel the need to report; "he feels competent to handle whatever it is." McGuire felt that it was "a little dangerous" to be away from Washington too long, since he wouldn't hear until he returned "what went wrong."

REORGANIZATION AND AUTONOMY

In 1905, jurisdiction over the forest reserves was transferred from the General Land Office in the Department of the Interior to the Forest Service in the Department of Agriculture. Since then there have been a series of proposals to return the forests and their administrative agency either to Interior or to a reorganized Interior agency, perhaps a Department of Natural Resources. To date, the Forest Service and its supporters have prevented such an occurrence. Three transfer attempts happened while McGuire was chief.

During his 1968 campaign, Richard Nixon promised to reorganize the federal bureaucracy to make it more governable and efficient. In 1970, the year President Nixon signed the National Environmental Policy Act, he began to implement recommendations from his Reorganization Advisory Committee, which included creation of the Environmental Protection Agency by executive order. Four programs in several departments that dealt with water quality, air pollution, pesticides registration, and radiation standards were merged to become EPA. He also converted the Bureau of Budget into the Office of Management and Budget, a step toward streamlining administration of the federal government. When the president turned his attention to natural resource programs, however, he ran into stiff resistance.

The Nixon proposal was ambitious—replacing seven cabinet-level departments with four, one being the Department of Natural Resources. Under this plan, the Forest Service and the Soil Conservation Service from Agriculture would be in DNR, along with the National Park Service and the Bureau of Land Management from Interior. Since the issue was debated at the departmental level, the proper place of the Forest Service and the other agencies received only modest attention.

OMB well captured the need for reorganization of the natural resource agencies with language that all chiefs from McGuire to Dale Bosworth could identify with: "Federal natural resource programs have developed on a piecemeal basis over the years, resulting in programs scattered among agencies with attendant overlaps, inefficiencies, and voids." The time was right to address jurisdictional conflict and cross purpose.

McGuire was concerned that this effort to place land management agencies in a single department would in fact cause a splitting up of the Forest Service by sending its three branches to different departments. National Forest Administration would move to DNR, Research to a new Department of Science and Education, and State and Private Forestry to a new Department of Economic Development. Thus, even though McGuire conceptually agreed that some sort of reorganization was in order, in fact long overdue, the Forest Service would lose too much.

The second reorganization effort, also during the Nixon administration, was more modest and called for a Department of Energy and Natural Resources. However, Congress was willing to support only energy matters; the natural resources portion of the proposal died in committee.

The third effort to transfer the Forest Service occurred during the Carter administration. Carter had campaigned against the bureaucracy and, like Nixon, had little congressional support. His proposal, too, failed, although he was successful in creating the Department of Energy.

To McGuire, the biggest hurdle for reorganization would be to deal with the much more politicized and centralized conditions in the Department of the Interior, compared with the Forest Service in Agriculture. He recounted his favorite story about power lines crossing the Rocky Mountains. "Where the line crossed national forests, they'd get a permit from the district ranger. When the line crossed Parks or BLM or some wildlife refuge, the permits were always signed by the secretary of the Interior." Also, since the Forest Service research program was of a size and quality very unusual for Interior agencies, there was a good chance that

Chief McGuire with Assistant Secretary of Agriculture Robert W. Long. USDA Forest Service photo.

it would be lost during the proposed merging with Interior. Added to that would be a loss of the link to agriculture and the Forest Service's important, historic ties to farm woodlots. These were only some of the elements in McGuire's equation on reorganization; altogether, the costs would be greater than the benefits. Looking back at his tenure as chief, McGuire was proud that he succeeded, "with the help of many others and with the help of circumstances, in keeping the Forest Service in the Department of Agriculture."

RENEWABLE RESOURCES PLANNING ACT

"Jim Giltmier, staffer for the Senate Agriculture Committee, decided we ought to have a little brainstorming back around 1973" on how to persuade Congress to take a longer view of forest management, and then make a dollar commitment that went beyond the current fiscal year. The brainstorming indeed took place, and McGuire recalled "speaking in favor of some kind of legislation that would require the Forest Service to submit long-range plans." In his interview, he listed five major program proposals since 1920 that had been difficult to achieve because, although the administration might agree that a proposal was valid, there was great reluctance to commit to future expenses. Thus, step 1 was to find a way to "justify our annual appropriation requests, because most forestry programs are multiyear." He continued, "We needed some way that would force the administration to release long-range program information." Under the Forest and Rangeland Renewable Resources Planning Act, known as RPA, the administration would be required to submit long-range assessments and programs. Then Congress, supposedly, would be influenced to be more generous and take the longer view.

The Area of Agreements Committee, hosted by the American Forestry Association, played a major role in RPA's enactment in 1974 by getting agreement from a diverse group of environmentalists, conservationists, and industrialists that the law would be of benefit to their special interests. "The only opposition came from OMB, which opposed it on the grounds that it subtracted from the president's authority." The bill cleared Congress "without too much trouble" and arrived at the White House "right at the time Nixon had left the Oval Office and Ford was sitting down. It must have been one of the first bills to arrive at Ford's desk." Accompanying the bill was a strong letter of endorsement from Secretary Butz and "an equally strong letter from OMB urging the president to veto it." McGuire believed that ordinarily Ford would have agreed with OMB but was reluctant to "veto the first thing he picked up."

RPA generated a great deal of useful information, and there is evidence that it did manage to loosen congressional purse strings a bit. However, the various reports were very controversial and enormously expensive to compile, and eventually it was clear that long-term commitments were impractical in a political arena that changed, sometimes abruptly, every two, four, or eight years. The Forest Service would continue to carefully plan for the future, but not via RPA.

NATIONAL FOREST MANAGEMENT ACT

In 1973, a federal judge ruled in *Izaak Walton League v. Butz* that the clearcutting on the national forests within his circuit were in violation of the 1897 Forest Service Organic Act. The lawsuit had been in response to clearcutting on the Monongahela National Forest in West Virginia. The judge recommended that instead of appealing his decision, the agency instead seek a legislative remedy. The result was the National Forest Management Act of 1976, which officially was an amendment to RPA.

It would seem that the rather detailed NFMA significantly reduced Forest Service discretion to make timber sales. However, McGuire pointed out "most of the direction in the Forest Management Act is direction to do what the Forest Service was already doing, so it was difficult to argue that that should not be put into the law." Although the agency would have preferred a simple bill, "there

Brock Evans, the Sierra Club's Washington, D.C., representative, meeting with Chief McGuire, 1973. Preserving wilderness and reducing clearcutting were important goals for Sierra Club members. USDA Forest Service photo.

was no chance of getting such a bill through Congress, because there were so many interests contending to put other things into the law." The process "came down to a search for middle ground with the debate continuing right into the final conference committee." Given the possibility of Ford's not being reelected and Secretary Butz's departure, the "situation was extremely fluid," and McGuire "never did succeed in getting an administration bill forwarded for Congress to consider."

NFMA mandated that the national forests be managed on a sustained-yield basis, apparently no change from the 1960 Multiple Use–Sustained Yield Act, except listing "wilderness" as one of the uses. However, the "issue of nondeclining even-flow as an interpretation of the sustained-yield mandate came up as a result of the Douglas-fir supply study." With the advent of computers in the 1960s, timber planners were able to "project future stand structure into the second rotation." Even more powerful computers allowed projections into the third and forth generations, "and it became apparent that cutting at the pace that was being tolerated would eventually result in decline in the allowable cut of Douglas-fir. It was obvious that to stay within sustained yield over long periods of time, we'd have to ensure that there was no such decline in the distant future." That insight prompted a manual directive on nondeclining even-flow policy.

Shortly the new policy began to affect allowable cut, and the timber industry "got stirred up." When the committee proposed to include nondeclining yield in NFMA, "it was difficult for the Forest Service to argue that it should not be done." McGuire did request that the language be "loose enough to provide for exceptions," such as the need to treat an insect outbreak. OMB probably would have joined the timber industry in opposing the language that would apparently reduce timber sale revenue sometime in the future. However, "we never asked them" what they thought. "We took the position that the law requires us to operate the national forests on a sustained-yield basis and that the law left it to us to define what that meant."

Clearcutting turned out not to be controversial because "the whole section came right out of the [Senator Frank] Church guidelines," which the agency had endorsed earlier and had helped draft, well before NFMA. However, what Congress elected to call "collusive practices obviation" became controversial after passage.

The 1897 Organic Act required that national forest timber be sold at public auction, and after more than seventy-five years of experience, the process and its strengths and weaknesses were very familiar to the agency and the many buyers. From time to time, buyers had engaged in collusive practices: one company would privately agree not to bid above a certain amount on a particular sale to allow another to obtain the timber at a lower price, and at the next auction, the roles would be reversed. Some in Congress wanted oral bids at auction, but most favored sealed bids. McGuire figured that "if Congress wanted us to use only sealed bids, we'd be glad to do it." When he testified, he did not object to the sealed bid requirement, but he did explain the pros and cons. No one wanted to speak out against the process—it seemed so logical—and it was included in the statute. "Once it was passed, all hell broke loose—the industry objected" and got to Senator Church. Promptly, Senators Church, John McClure, Mark Hatfield, and Robert Packwood met with McGuire in Packwood's office. "The four of them got me in there and told me that I'd have to abandon sealed bids. I said, I can't, you've just passed a law requiring them." They told McGuire to find a loophole; he responded that there were no loopholes. They "worked me over for about an hour," but the chief was not able to agree to ignore the requirement for sealed bids.Two years later, after much debate, Congress reinstated the oral bid option.

At the time NFMA was being debated, Max Peterson was deputy chief for Programs and Legislation and thus was much involved with the bill. "The enormous respect John had with the Hill allowed him to work with both sides pretty effectively in trying to fashion that bill," Peterson recalled. The Senate version included a detailed planning process, but the House version was much shorter and contained fewer restrictions. Which version to support "ended up being a fairly big question for the administration to resolve. So the department and OMB got involved, and we ended up having a meeting on the Hill. It was quite an acrimonious meeting, and finally John said, 'I think we need to follow the planning process, because if we don't, we won't get much discretion, we'll get a whole bunch of requirements in the bill and some federal judge downstream will decide what they mean.'" According to Peterson, McGuire also asked a pointed question: "Even if you think that's the best way of doing it today, what are you going to do when Research tells you something different?" Eventually there was agreement that the administration would support the planning section of the Senate's bill.

NFMA authorized a Committee of Scientists to advise the secretary, and the committee turned out to be a "mixed blessing," according to Peterson. The appointment of the committee was delayed eight months by the change of administrations from Jerry Ford to Jimmy Carter. The new assistant secretary of Agriculture—Rupert Cutler— "wanted to hear from people throughout the country." As a result, "they spent a long, long time developing those planning guidelines," and the process contained much, much more detail than anyone had envisioned.

The resulting regulations for NFMA required maintenance of viable populations throughout a planning area. By the time Dale Robertson was chief, this requirement, more than the Endangered Species Act, would have an enormous impact on Forest Service timber sales. Peterson remembered that as the bill was being revised, "we put in minimum populations of the native species throughout the area and were really feeling that this was a good idea. I don't think that anyone thought there was any potential for causing problems...The diversity of wildlife was not, in my memory, a real big issue that was debated."

Despite the importance and complexity of this particular legislative process, there were light moments. Peterson, sitting in for McGuire, recalled that during the House-Senate conference on NFMA, "the room was completely packed with people." When the committee broke for a vote, "a House staffer who had been out of town and apparently didn't realize what we had come to, as the way we were going to approach the bill, walked over in front of me and said, 'You're supposed to be following the administration's position.'" Promptly, someone from the White House, the department, and OMB "took up positions behind me in the conference room....Jim Giltmier from Senator Talmadge's staff spotted these people and realized I was in kind of a spot. He sent a note over that said the senator wants to know who those goons are." Peterson replied, "Don't worry about it; just try not to ask me very many opinion questions."

As NFMA was moving through Congress, another related bill was under consideration, the Federal Land Policy and Management Act, also known as the BLM Organic Act. By executive order in 1946, President Truman had created BLM by merging the General Land Office and the Grazing Service. The agency has responsibility for more than 270 million acres of federal land and 570 million acres of mineral rights. In contrast, the national forests comprise 191 million acres. Thirty years later, Congress set out to determine just what BLM was to do.

McGuire did not believe that the national forests were much involved with the proposed law, but "there were some very important parts of that act that affected us, such as mineral rights" and grazing. "We tried to get the committees to keep that bill separate from the Forest Service, but they kept writing in language that applied to all the public lands." He believed that "one of the landmark parts of the act" was in the findings section, where Congress found after two hundred years "that the better policy is to retain and not to dispose of the public lands. Just like that, so nonchalantly."

McGuire retired on June 15, 1979, and was succeeded by R. Max Peterson. In retirement, McGuire was concerned that he might appear to be taking advantage of having served as chief, and so he limited his forestry-related activities to occasional teaching assignments, serving on nonprofit boards of directors, and appearing on conference programs.

CHIEF R. MAX PETERSON

USDA FOREST SERVICE PHOTO

As chief, R. Max Peterson faced some of the same issues that had challenged his predecessors, such as clearcutting and wilderness. He also encountered new problems of great import: severe budget cuts, endangered species, job discrimination litigation, and even a spectacular volcanic eruption. But his career in the Forest Service included another earth-shattering event—a magnitude 7 at that.

Peterson had been born in Doniphan, Missouri, on July 25, 1927. Following military service during World War II, he entered the University of Missouri to study engineering. He graduated in 1949 with a degree in civil engineering and began his Forest Service career on the Plumas National Forest in California. Following stints on two other national forests in California, even though his retreat rights were uncertain, in 1958, Peterson took advantage of an opportunity to attend Harvard University to earn a master's degree in public administration. He would find that both the formal education and the relationships formed would be beneficial in the future. Too, the degree gave him a breadth of view just at the time the Forest Service was attaching importance to such things.

In 1959, he moved to Missoula, Montana, as assistant regional engineer. Shortly after his arrival, a severe earthquake tumbled the Madison River Valley near West

Source: An Interview with Ralph Max Peterson, *by Harold K. Steen. Durham, North Carolina: Forest History Society, revised 2002. Peterson was interviewed in 1991 and 1992.*

Yellowstone. Peterson remembered, "about thirty people were killed…and a whole bunch of them are buried under that slide. …It was international news. …It was a Mount St. Helens." The disaster brought extraordinary engineering challenges, and Peterson rose to the occasion. Decades later, some of his long-time associates urged the interviewer to include questions about Madison River, which they remembered as a significant event.

In 1966, following five years in the chief's office, Peterson was transferred to San Francisco as regional engineer for the California region. He liked where he was and what he was doing, but in 1971 he moved to Atlanta as deputy regional forester, abandoning his long-held engineering path. Within a year he was promoted to regional forester. The engineer had been neither a district ranger nor a forest supervisor, and now he was a regional forester, just one level below chief.

Chief McGuire had required his regional foresters to agree in writing that they would accept any offered assignment. Even though Peterson very much enjoyed life and work in Atlanta, when McGuire asked him to move to Washington, D.C., as deputy chief for Programs and Legislation, he really had no option but to comply. In fact, McGuire had been under great pressure from Congress to end the "foresters can do anything" syndrome, and Peterson's M.P.A. had given him the sort of credentials that were directly responsive to complaints from the Hill. He did well in his new assignment, and on July 1, 1979, Peterson succeeded McGuire as chief.

Peterson has a memory for detail and is a great storyteller, using the full range of body English and facial expressions to animate the tales he tells with such relish. As did McArdle, when Peterson was in the field to see firsthand what was going on, he took time from the organized tour to sneak off with local staff and enjoy their companionship, and maybe indulge in a bit of shoptalk. There are snapshots of the chief sitting at campfires, with a stub of a cigar in the middle of his wide grin, playing some sort of card game—five-card draw, perhaps—with ranger district and national forest personnel who also were enjoying the informal evening with the chief.

SELECTION AS CHIEF

Douglas Leisz was associate chief under McGuire, but Peterson had longer tenure in Washington as deputy chief. Both were well known and highly regarded in the agency and in Congress. Shortly after McGuire announced his intention to retire, he told Peterson that "they boiled down candidates for the chief's job to you and Doug Leisz." Secretary of Agriculture Robert Bergland wanted to interview both of them, calling Leisz back from an Alaska trip to do so. In fact, Deputy Secretary James Williams conducted the interviews, and Leisz returned to Alaska. A week passed, and Peterson chanced upon McGuire in the hall. "Looks like Bergland decided you're going to be chief, and he wants to see you." Peterson met with Bergland and Williams, who commented that it had been a "tough decision" and that they wished there had been two jobs instead of one. The three easily agreed that Leisz would remain as associate chief.

THE CHIEF'S DAY

Peterson remembered that agency personnel, whoever their preferences for chief might have been, quickly closed ranks in support of his appointment. And there was just so much to do that he ran "wide open" during his tenure. On page 1 of the daybooks he kept while chief, dated July 2, 1979, he noted a meeting about the Alaska oil pipeline, a message to the Committee of Scientists about the National Forest Management Act, a notation about a message from the Council on Environmental Quality, forest fires out West, and some Senior Executive Service positions that needed filling. And then he began roughing out his calendar. He had inherited meetings that McGuire had made commitments to attend—the National Forest Advisory Committee, the State and Private Forestry Advisory Committee, the International Association of Fish and Game Agencies, and the American Mining Congress. Peterson and Leisz sat down and decided who would attend which meeting. A busy first day indeed.

Peterson continued the brief and informal early-morning meetings with his deputies and associate deputies that McGuire had begun. He considered the meetings "enormous timesavers in terms of being sure that things are geared together." To that he added occasional executive sessions away from the office "to talk about where we had been and where we were going."

When the Reagan administration came in a year and a half after Peterson had been appointed chief, the position of assistant secretary of Agriculture remained vacant for four months. Incoming Secretary John Block, even before he was sworn in, gave Peterson formal delegation of authority to serve as acting assistant secretary, which included oversight for certain other Agriculture agencies, such as the Soil Conservation Service. Peterson worked mainly out of his Forest Service office, but once a day he walked across the causeway between South Agriculture Building and the department's headquarters. He would read his mail, sign letters, and "moved all of the stuff that needed to be moved." He played the dual role from January 19, 1981, until mid-May, when John Crowell was sworn in as assistant

Chief Peterson and Idaho Governor John Evans at a televised news conference discussing an escaped prescribed burn, 1979. Use of prescribed fire to improve forest "health" would become more and more common. USDA Forest Service photo.

secretary. Crowell came from a forest industry background and well understood the Forest Service. He would be a very significant player.

ROLE OF ASSOCIATE CHIEF

The Forest Service had operated without an associate chief from 1944, when Earle Clapp retired, until 1966, when Cliff's reorganization plan was adopted. During those two decades, assistant chiefs would rotate the responsibility for being acting chief whenever the chief himself was away from Washington, D.C. They had authority to sign most mail and in other ways ensure that day-to-day activities were on course. Even after 1966, deputy chiefs served as acting; however, by the time Peterson was chief, "the acting chief as matter of importance went out." The associate chief "has the same delegated authority as the chief, unless the chief withholds something from him." Only a few times during the year would both the associate chief and chief be out of town at the same time, and never out of the country at the same time.

Leisz was older than Peterson, and when he retired in 1982, Dale Robertson succeeded him as associate chief. Peterson wrote a memo to his new associate to help define their relationship. First, he said, the memo would be general, as "detailed written guidelines spelling out this relationship are probably counterproductive." He went on, "It is my intent that the associate chief continue to operate as an alter ego to the chief, so the decisions can be made promptly and effectively." Peterson said that he would rely on Robertson's "good judgments" to know when he needed to consult with the chief before making decisions that he was legally able to do. Areas where he expected consultation were nominations and reassignments to the Senior Executive Service, important budget decisions, and major legislation. Robertson would be expected to review SES employees' performance, review and approve the Washington office budget, and represent the agency in meetings with the Department of Agriculture and other federal agencies.

THE WHITE HOUSE AND THE DEPARTMENT

The "White House" is more than the president's home; to Washington insiders, it's also a fluid aggregate of presidential advisers and institutions. Depending upon the president's style, a bureau chief from time to time can be directly influenced or even directed by the president's economic advisers, science advisers, and more recently, environmental advisers. On occasion the Forest Service would send draft language to the White House "for them to consider putting in the presidential statement" on a particular subject. Too, the relationship between the president and the secretary of Agriculture was very important to the Forest Service; a strong friendship meant that the secretary could run his own show, but a distant acquaintance of the president might find that the White House had opened its own channels to the bureaucracy.

The highest-profile White House agency is the Office of Management and Budget. Peterson thought that OMB "has a multifaceted set of responsibilities which most people don't fully understand." He noted that "all positions on legislation must be cleared through OMB." During the Nixon, Ford, and Carter administrations, the Forest Service recommended positions on some of the fifteen hundred pieces of legislation that it tracked each year. Mostly, "nothing ever happened." But on occasion, the agency would "receive a request for a report on legislation," and perhaps testimony. The "rules require that a position on legislation can only be established by the administration, by a presidential appointee...You never say this is the Forest Service position"; it is always the department's or administration's position.

Rupert Cutler, Carter's assistant secretary of Agriculture, had involved himself mainly in wildlife and wilderness issues. Even though he was "highly environmentally oriented," Cutler "showed a lot of respect for professionalism in the Forest Service." Peterson pointed out that many staffers brought into Washington were field-oriented people; it was not unusual for them to spend so much time on the national forests that they spent inadequate time on budget preparation and related tasks. Cutler was especially critical of the performance of wildlife staffers in their preparation of budget materials. To Cutler, one of the reasons that the Forest Service did not fare well in the portion of its budget related to wildlife was weak staff work.

When John Crowell came in as assistant secretary during the Reagan administration, he usually cleared "the position on legislation...Generally we sent stuff to OMB at the same time we sent it to him." The Reagan administration ended the practice of including agency staff at OMB meetings that dealt with matters of budget or legislation. Peterson thought this change was "part of pulling the administration tighter together, and maybe for more candid discussions." As to Congress, if Peterson felt strongly against an administration policy, he would ask the secretary for relief—that he not be asked to testify in support of the policy. Pragmatism was part of Peterson's position; if he had testified one way last year under Carter, and then in a contrary way under Reagan, "you're going to be in a real spot then if they ask you about it. 'How come you changed your mind since last year?' You can't perjure yourself."

Crowell came from industry, the abrupt shift from Cutler reflecting contrasting presidential philosophies. Crowell told Peterson, "I don't intend to ask for you to step aside as chief,...but if you decide at some point you want to do that, that's your decision." Crowell went on, "I expect that we'll discuss things and we'll wrangle out decisions and sometimes disagree and once we make a decision we'll go forward." Peterson concurred. The Reagan administration had come in with "a very well-defined agenda to reduce the domestic budget and workforce," and the federal lands were "to be used more to meet a wide variety of needs." Crowell is remembered today mainly for the pressure he put on Peterson to increase the cut from national forests from thirteen billion board feet per year to more than twenty, a goal reminiscent of the McArdle years. Neither goal was even approximated.

The chief liked both assistant secretaries as people, and later when all three were no longer in government service, he joked to them that "it sure would have been a whole lot easier for me if they put you two guys in a sack and shook you up and gave me the average." As to assistant secretaries in general, Peterson thought that "if the government's in peril today, it's because there's been a tendency in the last twenty years to appoint assistant secretaries who have line authority over major agencies who have practically no background on the subject matter." Too, since they "don't have any tenure," they tend to go along with whatever the secretary wants so long as it's "not illegal." Finally, they were young, "usually in their thirties," and were there primarily to get a "bigger job somewhere else."

DECENTRALIZATION

When Gifford Pinchot received authority to manage the national forests in 1905, he stated, "Local questions will be decided on local grounds." This decentralized form of administration has remained a Forest Service hallmark and source of great pride. However, times changed and complexities increased, and each subsequent chief had to figure out for himself just how decentralized the agency could remain and still carry out its ever-expanding mandates. It was the same for Peterson, who noted "trends that run in both directions." As communications and travel improved and the public and others could better see how different national forests under basically the same conditions would treat the same operation in different ways—logging slash disposal and wilderness rescues were just two examples—questions would arise. Peterson sought "reasonable consistency" for practices on ranger districts within a national forest, and for national forests with a region. Failing that, then Congress might well get involved and say, "This is the way it's going to be done."

Peterson remembered that "Ed Cliff used to approve every timber management plan for a national forest." That sort of action could certainly be seen as part of a centralizing trend; Cliff had begun implementation of a plan to make the national forest the basic planning unit. In his day, a permit for a ski area needed approval at the regional level, for example, but a timber plan needed to go to the chief—certainly an "inconsistency." It was not until the National Forest Management Act of 1976 that the national forest "really became the planning unit." Peterson did not see this as centralization but as consistency.

"There is a potential for misuse of NFMA; for example, I would not let the policy staff get mixed up in a plan for national forests." There was a technical review of a plan's components by planning staff, but which alternatives to adopt was "off limits." The regional forester and not the chief was to make such decisions, "which was appealable to the chief." Peterson thought that "the only thing running contrary to that principle" was that Assistant Secretary Crowell and Douglas MacCleery, his deputy, "had a strong view that the western forests particularly ought to be able to produce more timber." Nonetheless, Crowell never looked at a forest plan and said, "I want you to pick this alternative." The assistant secretary got involved only when an appeal came in.

Peterson thought that overall agency structure had a great deal of influence on how decisions came to be made. For example, the size of ranger districts, the number of districts per forest, and the overall number of regions affected not only the chain of command but also a land manager's knowledge of local conditions. Then there were the four levels of management—Washington office, regional office, national forest, and district—that had been around since Pinchot, with all of the opportunities for duplication of decisions. Peterson thought that the Forest Service either had been "awful smart a long time ago" or was out of date now. "And it may be some of both."

WORKFORCE DIVERSITY

It began while McGuire was chief: a disgruntled applicant to the Pacific Southwest Forest and Range Experiment Station in Berkeley charged the agency with sex discrimination. She won. Ultimately, there was a class-action suit, and the judge handed down a consent decree mandating that the Forest Service in California hire women in all fields and at all levels comparable to the women-to-men ratio in that state's workforce.

The Forest Service receives legal advice from the Office of General Counsel in the Department of Agriculture. A general counsel had "looked at the numbers" and, in Peterson's words, decided that "it looks open and shut that the Forest Service has not really reached out to bring in women and promote women. You've been happy to have this all-male Forest Service." She added, "Your case is not good," and signing the decree seemed the correct course. Peterson felt that the Forest Service did not really understand the decree and agreed to its terms without first running the numbers and asking for clarification. The decree had come down during a time of full budgets, but it fell to Peterson to implement it while trimming the overall agency workforce by twenty-five percent after Reagan became president.

Peterson's hill got even steeper. He found general concurrence within the agency, by both men and women, that they had made "good-faith efforts" to comply with the decree. Especially, "the women knew that the Forest Service was doing about all it could, that people were making new progress, that people were being placed, people were being recruited." Enter the Department of Justice, which declared that the Reagan administration was opposed to quotas, which it held were illegal, and the decree was based upon quotas. The Forest Service wanted to extend the decree to obtain more time to fulfill its intent, but

Justice, Peterson recalled, replied that "there isn't any way that we are going to agree" to an extension. Instead, the agency should document its progress under the decree and stop.

The attorney representing the class-action suit that had yielded the decree learned of the new strategy and decided to "play hardball" himself. He went back to court to have the secretary of Agriculture "held in contempt." Secretary Richard Lyng was civil rights officer for the department and "was very much in favor of a diverse organization," Peterson said. He made each agency head "personally responsible for taking action" to hire more women and minorities. Peterson reported to the secretary, giving him background on the decree, and explaining just how the Forest Service had been in error for not having read the decree more carefully. Lyng responded that "people are dinging on" the president for "civil rights things." It was Lyng who would have to take the heat.

With the issue now well above the agency level, the Department of Justice reversed itself, announcing that "we've got to take all of the action we can to prevent the secretary from being cited." By that time, however, "the wells had been so poisoned" that nobody would believe that the Forest Service was guided by honest motives. "The judge did in fact hold the secretary in contempt and ordered the extension of the consent decree for several more years." To Peterson, "the implication of the Forest Service ended up being twice as bad as if we'd gone ahead and worked out the thing as we were about to do." And, he remembered, it could all have been avoided.

BELOW-COST TIMBER SALES

During the 1950s, as the timber cut from national forests was reaching a third of timber cut from all lands, the Forest Service began promoting the idea—inside and outside the agency—that it "made a profit." Management of the forests did not cost the taxpayers anything because timber sale receipts, which were revenue to the general treasury, were larger that the amount that Congress appropriated to fund the entire agency. Looking at the numbers from a distance, it is not clear whether the Forest Service ever made a profit, but it did indeed generate a great deal of revenue in relation to its budget. However, within the agency at the time, its "profit" was a "fact" and a source of great pride. As time went along, the costs increased and the revenue did not, but the Forest Service continued to believe in its power to generate revenue through sales. Congress was also supportive. However, with the advent of public opposition to clearcutting and ultimately all logging, there would be more and more challenges to the Forest Service accounting system that showed a profit. Eventually, by the time of McGuire's and Peterson's administrations, the label "below-cost timber sales" would be used by those who asserted that logging in the national forests not only was not profitable, but in fact was being subsidized by tax dollars. Plus, it was ugly.

Peterson "didn't think below-cost timber sales is the issue. It never has been the issue." The issue has "really been a question of whether national forests ought to be primarily used for other purposes." Accounting practices were part of the question; must a Forest Service activity be operated in the black each and every year? If so, for example, a precommercial thinning of a young stand to release selected trees—the poles cut are too small to be sold to defray the full costs—would be a deficit operation.

An important part of the equation for Peterson was the "enormous increase in the cost of preparing sales nowadays." Detailed analyses of site conditions by a variety of disciplines and preparation of impact statements had been central to cost increases, and Peterson thought that "more money is being spent on the documentation in some cases than it's worth." Most of the so-called below-cost sales were in the eastern portion of the country in relatively young stands that required investments to yield a mix of values. Instead, most of the logging controversy was in the Pacific Northwest, where the sales were clearly "profitable." The real question for Peterson was, "How do you evaluate management alternatives for forests?"

TIMBER SALE BAILOUT

Another controversy over the profitability of Forest Service timber sales began in the late 1970s. It was a time of high inflation; interest rates for thirty-year home mortgages ranged between fifteen and twenty percent. It was also a time of speculation by the timber industry, mainly in the Pacific Northwest, as it bid on Forest Service timber sales. With high inflation, a company could comfortably bid "too high," knowing that during the three-year life of the sale, inflation-driven price increases would more than cover its bid. Then "the bottom fell out of the housing market and the price of lumber," Peterson recalled, and companies found themselves contractually bound to pay for timber that was suddenly "overpriced."

Peterson's "basic feeling was you signed a contract, nobody forced you to sign a contract, and you've been making a lot of money recently." The chief thought, "You've made your bed, and you're going to lie in it." Assistant Secretary Crowell was "highly indignant" at industry complaints, as he was a strong believer in the free-market system, and "you live and die by the free market." But it was not that simple; companies had made only a minimal deposit and had only promised to buy trees at a certain price. It seemed highly unlikely that the government would ever be able to collect the promised money, especially

Associate Chief George Leonard
Timber Sale Contracts

In the early 1970s, there was a significant recession in the timber industry. The Forest Service looked at the situation and routinely provided some extensions on timber sales. Instructions went out from the Washington office that sales that had met certain criteria in terms of their prices could be extended for an additional year. I think that was done first in 1973.

When the market didn't come back in 1974, we authorized some additional extensions, and then in 1975 I had become the assistant director of timber management for sales. I routinely—on my own authority—authorized additional extension of sales in 1975. By 1976, the market was good, and we quit it.

Then we got into the 1980s, during the Reagan administration, and John Crowell was the assistant secretary. John's objective was to get the timber harvest levels up on the national forests, but he was frustrated by a significant recession in the first two years of the administration. Then the market was good, but timber supplies on private lands were restricted as a result of cutting all the way back into World War II. So the ability to increase harvest and respond to the good markets pretty much was limited to public timberlands. There was extreme bidding on the assumption that the markets were going to be better next year, not on the basis of what they could pay this year.

Two things had happened that made things a little different. First, in the late 1970s, because of the concern over bidding, we had made a number of changes in the timber sale contract, trying to restrict extensions. There was not direct authority within the timber sale contract just to take that easy out and give them an additional extension. When the problem first came up—when I was prime mover because I was either in sales or director of timber management—we said, No, we are not going to give extensions.

The industry immediately went to Congress and sought legislation to authorize those extensions. Immediately, because positions on legislation have to be taken by the department, not by the agency, the department got involved. Initially, Crowell opposed legislative extensions, partly because it was inconsistent with his objective of getting harvest levels up.

There was real concern within the administration. We went through one legislative session, and nothing happened. So then internally, within the agency, we developed a program of extensions that became known as the Multi-Sale Extension Plan. We said we won't treat individual sales and just give you an extension on an individual sale. If you've got problems, you've got to come in with all the sales you have under contract and give us a plan for operating all those sales. We'll give the approval for the plan as a whole, rather than on individual sales. We will impose

continued on next page

continued from previous page

requirements that you pay so much each year, whether or not you harvest. If you want five years of extensions to get this group of sales logged, you're going to pay twenty percent per year, even though you may not harvest. You've got to pay up front.

We got the department to approve this program of multisale extensions. We got a whole bunch of those approved and authorized, and because of the tightening that we had done on the timber sale contract, rather than just authorizing those extensions, as we had done back in the early 1970s, we actually had to make a change in the secretary's regs in order to implement that. So it took us longer, and some people in the timber industry became frustrated over the amount of time.

They went to Congress and ultimately got legislation passed to enable timber companies to buy out of the contract and turn the volume back to us. So we had a multisale extension plan, and then this congressionally mandated buyout program that became interrelated. One of the interesting things was, whereas in 1975 as assistant director of timber, I was able to authorize additional extensions of sales, but about 1982 with the Multi-Sale Extension Plan, the actual decision on that was made by the president himself. That shows how the Forest Service had begun to lose control of the day-to-day operational decisions, and it had escalated up, with Ronald Reagan's initials on the decision letter to authorize the multisales.

The White House became very much involved. Crowell actually ended up being opposed to the extension plan, and Secretary of Agriculture Lyng overruled him. In fact, Attorney General Ed Meese called me to get the details of how we were going to ensure that the industry performed on these contracts, if the president authorized us to go ahead: What were the mechanics we were going to follow to ensure that the companies didn't just sit around for a while and then ultimately default on the sales? That's where we got into that twenty percent payment each year. It was very much a decision at the White House level.

From An Interview with George M. Leonard, *by Harold K. Steen. Durham, North Carolina: Forest History Society, 1999.*

if a "company went broke." Enforced bankruptcy, with all of the other creditors in line ahead of the Forest Service, "would have gotten a bunch of old worn-out equipment" that probably would not even run. Peterson decided that it "would be better to figure out some way that the companies could continue to operate" the timber sales. Too, they needed to "look at adjusting the price." Then, the industry began lobbying for "a big timber sale bailout."

Peterson recalled "all kinds of bills in Congress," most of them "too liberal" for the Reagan administration. The bill finally passed "ended up being a tremendously complicated formula," and "the question was, Was the president going to sign the bill?" Secretary Block and Assistant Secretary Crowell—Peterson was not included—"went to a meeting at OMB, it may have been the White House." They carried two briefing papers, "both on one page," prepared by the Forest Service, that summarized arguments in support of both signing and vetoing. Block supported signing, but Crowell thought that "the president ought to veto it." OMB Director Dean Stockman agreed with Crowell, suggesting that instead the Forest Service negotiate a settlement, company by company. However, the group could not figure out under just what authority or by which guidelines the Forest Service could proceed with such negotiations. Too, President Reagan was to be in Oregon with Senator Mark Hatfield the next day "right in the middle of the big problem area"; they decided that the president should sign the bill.

Although many people thought that the bailout was costly for the taxpayers, Peterson believed that "the record would be pretty clear that the timber sale buyout made money for the government." Under the bill, the sales were operated, bringing revenue instead of bankruptcy conditions, which would require the Forest Service to bear the cost of redoing the sales for remaining companies, plus the very high social cost of local industries' shutting down. In an imperfect world, Peterson believed, you "do what you do when you are dealt a hand that is not very good." The bailout was the "lesser of two evils." A significant lesson for the agency was, since high inflation was the root of the problem, "if the Forest Service sees another inflationary spiral going, it better certainly take another look at how it sells timber."

FOREST SERVICE–BLM LAND EXCHANGE

When most of the western national forests were created, via presidential proclamation, no detailed or accurate maps were available, and a substantial amount of land not well suited or well located for national forest management was included within the gross boundaries of a new forest. In contrast, lands managed by the Bureau of Land Management were mainly what was left of the public domain after national forests, national parks, wildlife refuges, national monuments, and military and Indian reservations had been created, plus homesteads to settlers and grants to states and railroads. It was clear to anyone who looked at modern maps that the nation would be well served if some national forest lands were transferred to BLM and some BLM lands transferred to the Forest Service. Hence the Forest Service "had a long-standing program to interchange land with BLM."

Peterson remembered that during a hearing, a Montana senator said, "That big cloud of dust that you see in eastern Montana every morning is the Forest Service and BLM people driving past each other to go to their land." The representative also said that they were duplicating costs with offices and related expenses. BLM Director Robert Burford agreed that "it was nonsense to have this land pattern," that it didn't make any sense to the users, to the agencies, or to the local counties. Peterson then asked a couple of his staff to prepare maps that showed logical areas for transfer.

Concurrently, within the Reagan administration there was a campaign under way to reduce the federal deficit—asset management—by selling portions of the federal estate. "Out of the blue" OMB was also looking for ways to save money and roughed out an exchange of its own. "They were talking about an interchange way beyond anything I had envisioned" that would supposedly result in "massive" savings. Peterson was vacationing over Christmas, and Associate Chief Robertson reached him by phone, reporting that the "whole weight of the president" is behind the OMB proposal.

The OMB exchange would be announced in three weeks as part of the president's budget; in the meantime, the proposal was confidential and thus the Forest Service could not engage in advance briefing. "This provided us with no opportunity to brief anybody on what we're doing, we don't have the maps, we don't have the data, we don't have anything." Nonetheless, shortly the chief would be testifying to Congress, and he decided to talk with BLM Director Burford. The two agreed to appeal to their departments to pull interchange out of the budget so that they could work out the details "at a more leisurely rate," but they had no success. When Peterson and Burford testified, "we couldn't really give good examples of the size of it, the shape of the land, and so on." In Congress there was substantial opposition to the administration's proposal, where "efficiency" was seen as a euphemism for closing local offices and related job loss, a situation rarely supported by members of Congress whose districts would be affected. That proposal died, and the agencies went back to work.

Even so, there was a substantial difference between the Forest Service and the BLM. "Burford, who drew the first map, really scared the heck out of me. He

Chief Peterson with former Chiefs McGuire, Cliff, and McArdle, with Assistant Secretary of Agriculture John Crowell, 1983. USDA Forest Service photo.

was really remaking a map of the United States. I mean, he drew a map and said everything in the East goes to the Forest Service, and then he drew a line sort of down the continental divide in Colorado and said everything here goes to BLM and everything there is national forest.…He said if were going to do this, we ought to do something that really does something, we ought not do anything half-hearted."

Even tougher to manage were the differences within the two agencies, whose staffs could support additions but not subtractions. Peterson invited former Chiefs McGuire and Cliff to come in and comment on the maps that were nearly final, before submitting them to Congress as part of a legislative package. All BLM land in Washington would go to the Forest Service "because there wasn't that much of it." In Oregon, "we would have picked up a big hunk of the O&C [revested Oregon and California Railroad land grant] lands, and picked up a lot of stuff in eastern Oregon." There would be "just a little bit of national forest land going the other way." Cliff responded, "Oh, this looks good." However, in northern California, the national forests in the Warner Mountains would go to BLM, whose holdings surrounded them. Cliff asked, "What happened to the Warners?" Peterson noted that "we worked for months to finally come up with an interchange proposal that probably in the final analysis nobody was perfectly satisfied with."

Included with the 1986 congressional package was a legislative environmental impact statement, which differed from a regular impact statement in that there was no public comment period. As stated in the two-hundred-fifty-page LEIS, the aim was to provide Congress with adequate information to prepare the legislation required to carry out the proposed interchange of "lands and minerals management responsibilities." The interchange was necessary to "resolve overlapping and intermixed BLM and Forest Service jurisdictions. The lands administered by the two agencies are similar, sharing similar users, management problems, and resources. In seventy-one communities, both agencies have offices with separate staffs performing similar duties." The LEIS listed a preferred action and several alternatives. The preferred action proposed transfer of 24 million acres of land between the two agencies and 204 million acres of mineral jurisdiction from BLM to the Forest Service. Three hundred fifty jobs would be eliminated, saving thirteen million dollars annually.

Despite the large effort by both agencies and the detailed recommendations with options, the proposal died in Congress. Peterson noted that people in Congress "were getting letters from home saying, Don't do this." He added, "We spent a lot of money on it. It was well supported in the administration by both secretaries." Finally, "The interchange was just too big, it was really too much." In a personal communication, Dale Robertson, who was associate chief during the interchange discussions, stated, "The one positive outcome was that the Forest Service got the authority to make surface-related decisions on mining projects on national forests. BLM retained 'down hole' authorities dealing with production and royalty payments. In my opinion, this was worth the whole effort by clearing up authority and responsibility for surface management of the national forests."

Former Associate Chief George Leonard, also in a personal communication, cited three reasons for the interchange failure. First, the two agencies tried too hard to balance local land swaps, "even though in many instances a major shift toward one agency or the other would have made sense." Second, user groups, such as the oil and gas industry, "tended to favor the status quo." It was a matter of record that the Forest Service was "not as committed" to mineral leasing as was BLM. And third, the formulas for payment to counties as a share of receipts were substantially different for the two agencies, with BLM distributing a full half of the revenue to the counties, but the Forest Service only a quarter. As so often happens in Congress, Leonard concluded, the interchange failed because "there was simply no significant constituency for governmental efficiency, [but there] were strong constituencies for maintaining the status quo."

Chief Peterson fishing on the Salmon River with Secretary of Agriculture John Block, 1984. USDA Forest Service photo.

Plaques certifying significant participation in the interchange effort were distributed within both agencies. One hangs on the wall of Peterson's home office. There was even a necktie. On the front was stenciled "Interchange." On the reverse, which could be flashed for an appreciative viewer, it read, "Bullshit."

WILDERNESS MANAGEMENT

Chief Cliff had initiated the Roadless Area Review and Evaluation program to study national forest roadless areas and their suitability as wilderness. McGuire was still chief—Peterson then was deputy chief for Programs and Legislation—when Assistant Secretary Cutler determined that the effort was "one step short" of actually yielding additional wilderness. Cutler thought the whole process needed redoing, and the next process was labeled RARE II. At the same time, the earlier study was redubbed RARE I.

Peterson recalled that RARE II "was quite a bit more ambitious." It not only involved roadless areas, it included "some areas that had not been previously inventoried that were considered nonqualifying because of roads or evidence of other development." In particular, RARE II "picked up new lands in the East that had been previously cutover, eroded, or farmed."

The RARE II process was completed about the time Peterson became chief, and President Carter was directly involved in making wilderness selections. Presidential assistant Stuart Eizenstat worked through the options with OMB Director James McIntyre and Agriculture Secretary Bergland, and together they recommended wilderness areas for the president to consider. "There was maybe a fairly naive assumption that after we completed the RARE II analysis, Congress would handle this as one big package...and we will somehow get through the big controversy over wilderness. That proved to be a wistful idea."

One issue was a court decision that found the RARE II impact statement inadequate. "This meant that if we were going to sew up some of the wilderness questions, we were going to have to do it through individual points and plans." The specifics of release language—describing lands that would not be designated as wilderness—was also vexing, which triggered "shuttle diplomacy between House and Senate, warring factions trying to work out something" about release language. The factions finally agreed, "if we designated certain lands as wilderness, the remaining roadless areas would be non-wilderness for at least one generation of planning." Then they could look at "everything again." Because of that congressional agreement, "we actually had the greatest movement...to the wilderness question...we'd ever had in history."

With the allocation "sort of behind us," Peterson felt that they needed to concentrate on management; after all, now twenty percent of the National Forest System had been designated as wilderness. The problems were varied; sometimes wilderness campers congregated in certain spots, causing deterioration. Too, there was no consensus among prowilderness groups just what to do to protect wilderness values; some wanted elaborate facilities and comfort. In addition, federal and local agencies, including the Forest Service itself, wanted to install things in wilderness areas: the U.S. Air Force wanted to place a radar tower, a county sheriff wanted a repeater to relay emergency radio transmissions.

Even the Environmental Protection Agency wanted to evade wilderness constraints. Aircraft could legally land in wilderness only under emergency conditions, but EPA wanted to land helicopters on lakes in certain areas to take water samples for its study of acid rain. When EPA came to the Forest Service saying that they were going to take the samples, "our people handling this said, 'No, you really can't do this without special permission, and it's unlikely that we'd grant such permission.'" EPA responded that the president wanted it done, to which the Forest Service replied, "Where's your NEPA analysis?" It quickly became

Gila National Forest Supervisor Bobby Williamson, Regional Forester Jean Hassell, and Chief Peterson, Gila Wilderness Area, 1985. The Gila was the nation's first, established in 1924. USDA Forest Service photo.

clear that "EPA had never made a NEPA statement on anything like this; didn't even know how to begin."

Forest Service wilderness staffers bucked the issue to the chief, and Peterson learned that EPA had a "tight deadline" for its national survey of water quality. When he asked why they could not use ground measurements for the wilderness sites, EPA responded that all of their protocols assumed helicopter sampling, and the whole survey could be "screwed up" if the Forest Service would not let them go ahead.

They "spent days and days and days" debating helicopter versus ground measurements and whether EPA needed to comply with NEPA. It became obvious that EPA then aimed to go over Peterson's head, and so he briefed Deputy Secretary Peter Meyers on the issue and EPA's plans to "roll us." Meyers met with EPA officials, telling them that the Forest Service "was on pretty solid ground and he was not inclined to overrule us unless they could prove their case. That put them in a whole lot better frame of mind for negotiations."

Negotiations between the two agencies continued, and they seemed to agree that they would jointly sample, by ground and by air, selected locations to determine whether there was in fact any difference in the data gathered both ways. "When we were really down looking at each other eyeball to eyeball," EPA again said that it lacked the time for ground measurements. To top it off, the president had scheduled meetings with Canadians, who were much concerned about acid rain. Peterson then asked how much money EPA had budgeted for the helicopter survey. When he heard the amount, "I said we'll take the money and do the job for you." As it turned out, the only sampling problems were caused by weather delays that affected the helicopters; the ground surveys were completed on schedule. Then, "we had a big celebration with EPA," and everyone was proud that the important project "had been carried off" and at the same time wilderness values had been respected.

MOUNT ST. HELENS

Peterson saw Mount St. Helens as providing "quite a test of the local Forest Service and other agencies' capacity of handling a major emergency." The volcanic cone is on the Gifford Pinchot National Forest, whose southern boundary is the Columbia River. After several false starts, the volcano erupted at 8:32 A.M. on May 18, 1980; a secondary eruption followed in June. The initial blast released enormous mud flows, and within minutes, tributaries to the Columbia were above flood stage as a wall of mud and water surged downstream. The blast leveled many thousands of acres of national forest timber and stands of the Weyerhaeuser Company. Fortunately, loss of human life was small, in part because of warnings and evacuations ahead of the blast.

Coincidentally on a western trip, Peterson had toured the area just days before the eruption. Washington State Governor Dixie Lee Ray had already closed the area to public use. As they flew over the steaming volcano, Peterson asked Forest Supervisor Robert Tokarczyk about a contingency plan "if the thing blew up." The chief placed a copy of the plan in his briefcase and returned to Washington, D.C.

On the day of the eruption, Peterson was in Georgia, planning to return home in the evening. His wife said that he'd had an urgent call from Robert Lake, public information officer for the Forest Service. She asked, "Did you know that Mount St. Helens blew up?" That was his first news, and he pulled the contingency plan from the briefcase. He called the Gifford Pinchot National Forest and "got right through...and was told exactly what was happening" with rescues and related activities. An unexpected aftermath was volcanic ash that

Mount St.Helens erupting on the Gifford Pinchot National Forest in 1980. The cataclysm tested state and federal agencies' emergency response capabilities. U.S. Geological Survey photo.

hampered rescue vehicles by clogging air-intakes in their engines, buried farmlands and city streets, and closed the Portland airport. Then there was an electrical storm that started forest fires throughout the area. "It was about as near as you could visualize what hell might be like." The physical facts were daunting enough, but "the real question was how are you going to deal with all the fear that was being churned out."

The next morning Peterson called the department to confirm that the volcano was on a national forest and to report that initial emergency operations were functioning as planned. He also said that "there are lots and lots of people that are really scared out there." The first response was from OMB: "For God's sake don't play this thing up because it's likely to end up costing a lot of money." Peterson needed to "play this down." OMB was concerned about "demands for money to dredge the river, to provide housing, an open bonanza."

Peterson thought that President Carter ought to visit the area and "wave the flag." He called his usual White House contact, suggesting just such a presidential trip. "The guy said, 'I don't know about this; well, I guess we can talk about it.'" Peterson left his office for a meeting with plans to go on to the airport and return to the St. Helens area. Before leaving Washington, he called his office and learned that he had just missed a ride in Air Force One; the president was already en route, but the chief could fly commercial and join him for his tour of the site. Carter did get very much involved "in waving the flag and showing concern." The president described the scene as looking like the "craters of the moon."

Some months before the eruption, Peterson had recommended to Secretary Lyng that the Mount St. Helens area be designated as a geological area, and the Forest Service had been exchanging land within the immediate area of the mountain to block up federal ownership. After the eruption, the urgent need to begin restoration of forest roads and bridges and to salvage thousands of acres of blown-down trees on both federal and private ownerships meant that quick decisions had to be made: how much to salvage and how much to leave to take advantage of an "unprecedented opportunity to track the recovery of the area." The Forest Service proposal ended up in Congress, which "went somewhat beyond the areas we recommended—but they stayed pretty close to it"—and created the Mount St. Helens National Volcanic Monument.

ALASKA

There are two national forests in Alaska, the Tongass and the much smaller Chugach. For decades the Forest Service had wrestled with these relatively remote areas that needed some sort of special treatment, such as fifty-year contracts to supply wood to local mills. But there were broader issues too; since attaining statehood in 1959 and with subsequent legislation that was transferring vast tracts from federal to state and native peoples' ownership, Alaska became a contest between saving natural conditions and encouraging development.

As part of this contest, the Forest Service had developed proposals to create major new national forests. While still deputy chief, Peterson toured the proposed forest areas and felt that the agency had made "pretty good progress" in getting the national forests established. When the Carter administration came in, Chief Peterson again toured the Alaska forests, this time with Assistant Secretary Cutler, who "was pretty well convinced that a lot of those ought to be national forests." After returning to Washington, Cutler made "an impassioned plea" for the new forests at a meeting at OMB. Peterson and Cutler learned shortly after that they had been "trying to hail a train and the train had already left the station." Secretary of the Interior Cecil Andrus had already obtained a commitment from President Carter to instead create national parks, national park preserves, and wildlife refuges. Thus, when Cutler testified to Congress, he was obliged to support the administration's position in favor of reservations to be managed by

Interior agencies. "When the dust all settled, the Forest Service got a little bit of land added, but not anything of any consequence." It could have been worse; for a while it looked as though the agency would lose land in Alaska.

There was more unpleasantness for the Forest Service. By the time the Alaska Native Interest Land Conservation Act became law in 1980, planning under NFMA was "way downstream" with published maps and statements about just how the two forests would be managed. Wilderness was the gut issue; not only would there be more wilderness on the Tongass, but under ANILCA, land of commercial value was being placed under native ownership, and the new owners pledged "a very high level of stewardship." The issue in Congress was Alaska jobs that would be lost, and Forest Service planners found themselves obliged to consider their proposed wilderness designations in conjunction with what was happening outside the national forests.

The local forest industry insisted that it needed 650 million board feet annually to operate its mills, and it was decided that the national forests would supply 450 million of that amount. However, to produce that quantity, the size of proposed wilderness on the Tongass would have to be reduced. Since shrinking the wilderness was not a viable option, the solution was to obtain a federal subsidy to fund access roads and reforestation—a tough proposal to sell to OMB. The proposal basically guaranteed a "balance between wilderness and jobs... .Meantime, the Carter administration had been defeated and suddenly we had the Reagan administration coming in." The best course was to get a bill through while Carter was still president.

"It was a terrible professional strain for the Forest Service to handle." The administration was going "the opposite way" from the Forest Service's direction, "but we still had to function within the system." In the end, Congress "in essence adopted the Tongass Land Management Plan into law." Alaska, however, would continue to present tough problems for future chiefs.

THE SPOTTED OWL

The Bureau of Land Management, like the Forest Service, was obliged to develop a management plan to ensure survival of the northern spotted owl. In Oregon especially, national forests were intermingled with BLM areas, and so coordination of plans seemed logical to Peterson. However, efforts to develop a joint plan "didn't get very far." Deputy Assistant Secretary of the Interior James Casons accused the Forest Service of "using a lot of poor information," and anyway, BLM lands "were different," and they had no need to "worry about this business."

Casons appointed a team of biologists to study the spotted owl situation, to look especially "for information gaps where we might not know as much as we should." Not only did the team of biologists conclude that current BLM practices would place the owl in jeopardy, they also recommended that the agency join the Forest Service in its efforts to produce an acceptable management plan. Casons was "enraged" by the conclusions and ordered "all the copies of the report turned in" for destruction or impoundment. He then "ordered the team to go back and change the report." Despite Cason's efforts to quash the report, a copy was made public. Later, when Peterson was no longer chief but was executive director of the International Association of Fish and Game Agencies, he opposed Cason's appointment as assistant secretary of Agriculture.

Peterson pointed out that over time, the spotted owl had been "marching backwards" onto federal land, as old-growth forests were being removed from private holdings. Thus, by the time the owl appeared to be in jeopardy, "the only place you can make a stand is on national forests and other public land." One thing was very clear: understanding of old-growth as habitat—for spotted owls and other species—was inadequate. In situations like this, advocacy groups could assert that the owl was threatened, or that it was not, and there was no good evidence to refute either claim. It had been Peterson's experience that people "tended not to get interested in research until you have a crisis," and even though Forest Service scientists had been studying the spotted owl for more than a decade, the new crisis spawned accelerated research.

The Forest Service had been moving toward a management plan "based on the best information we had at the time." The thinking was that it would be more effective to reserve large blocks of old-growth forest, perhaps twenty or so, rather than smaller, individual blocks for each pair of owls. When the Forest Service published its supplement to the impact statement, there was a challenge, and the department decided that the new policy went "too far." Deputy Assistant Secretary Douglas MacCleery "issued a decision that we should take another look, and we did." But when they looked at the owl management plan, "it came out that we needed more protection instead of less."

Even though some commentators insist that the issue was not "jobs versus owls," to many who were directly affected, it was indeed that simple. Forest industry-related jobs in the Pacific Northwest had already declined by twenty-five thousand because of more efficient technology and other adjustments during a changing economy. New wilderness areas and soft housing markets were seen as causing part of this decline, and now on top of all that came the spotted owl. Peterson said he didn't know whether more effort could have prevented the "train wreck" that occurred, but he remembered that the forest industry "roundly criticized every move we made to protect the spotted owl." Whatever the Forest Service did, it was "too much."

Peterson knew that Secretary of Agriculture Lyng "was asked to fire me several times by industry because I was too environmentally oriented, that I had put too much protection on the spotted owl." Lyng would ask Peterson, "Why are you doing this?" When the chief explained about indicator species and what they showed about old-growth forests, the secretary accepted his view. In fact, during a public meeting in the Pacific Northwest, he defended Peterson's actions. Looking back, Peterson could see that if they "could have gotten BLM to join in the overall conservation strategy...we could have pushed research along further." Then the "outcome of this could have been less of a crisis...But it turned out not to be."

Peterson retired from the Forest Service on February 2, 1987. He had picked his retirement date "at a point in a new administration where things are going reasonably well and where you've got two or three people that are good candidates to succeed you." That process would allow his successor time to become well established before the administration changed again. Secretary Lyng assured him that he could stay on as long as he wished, "that there isn't anybody here in the department that's thinking that you should retire." Peterson agreed to be named as chief emeritus and to travel on behalf of the agency for training purposes. Shortly into retirement, he accepted an appointment as executive director of the International Association of Fish and Wildlife Agencies.

CHIEF F. DALE ROBERTSON

USDA FOREST SERVICE PHOTO

While on leave from the Forest Service to complete a master's degree in public administration at American University, the future twelfth chief of the Forest Service found the campus "in an uproar over Vietnam." The students, many of them younger than this junior agency staffer, were outspoken about the war and whatever else was wrong with the "establishment." When Herbert Kaufman's *The Forest Ranger* was an assigned reading for a class, they believed that the book showed just how "militaristic" the Forest Service was. In only a few years, students like these—people who routinely challenged convention—would be hired by the Forest Service, and F. Dale Robertson recalled his campus experience as "a turning point." As chief, he would work to balance program management with the need for creativity.

Robertson was born in Denmark, Arkansas, on July 17, 1940. After he earned a bachelor of science degree in forestry from the University of Arkansas in 1961, he began his career with the Forest Service on the Deschutes National Forest in Oregon. A year later he moved to Washington, D.C., as a management analyst trainee, where he became acquainted with senior Forest Service staff who would remember him later. After two years in Washington, he moved to the Sabine

....

Source: An Interview with F. Dale Robertson, *by Harold K. Steen. Durham, North Carolina: Forest History Society, revised 2000. Robertson was interviewed in 1999.*

National Forest in Texas as an assistant district ranger. In 1966, Robertson was appointed ranger on the Choctaw District of the Ouachita National Forest in Oklahoma. He became the first chief to have been a ranger.

In 1968, Robertson returned to Washington as a management analyst, a substantial promotion from district ranger. "It was a great job because you got mixed up in everything in the Forest Service," he recalled. One of his assignments was to determine personnel ceilings, something that Congress and the Office of Management and Budget were routinely interested in. The difficult task placed him as a relatively junior staffer in direct contact with deputy chiefs to work out just how many personnel they could have, "but I survived all that. That was a test of a person, a young guy."

After four years as a management analyst, Robertson applied to have his name placed on the roster for forest supervisors. In 1974, he moved to Corvallis, Oregon, as supervisor of the Siuslaw National Forest, and after two years he moved to Portland as supervisor of the Mount Hood. His experience on those big-timber forests, at the time when clearcutting was more and more in the spotlight, convinced him that "we better stop some of this." His position was sorely challenged, and he realized that his "career could have gone down the tubes." But it did not, and later, as chief, he announced that clearcutting would no longer be a "standard practice" on national forests.

Robertson returned to Washington, D.C., in 1980 as special assistant to the deputy chief for Programs and Legislation. His primary assignment was to figure out how all of the many Resources Planning Act pieces could better fit together. The following year he moved up to associate deputy and became a member of the Senior Executive Service, as well as a full member of the chief and staff meetings. One of his tasks as associate deputy was to put together necessary legislation to create the Mount St. Helens National Volcanic Monument; "It was the issue I cut my teeth on, on how you take an issue and work it through the poky process in Washington." It also got him better acquainted with Secretary Richard Lyng and Assistant Secretary John Crowell, who favored a monument but with minimal boundaries so that vast areas of blown-down timber could be salvaged. As Robertson briefed Lyng on the issue, the secretary turned to Crowell, saying, "I think you are way too conservative with those boundaries."

Robertson served well as a role model—a prototype, if you will—for others who were more and more being "fast-tracked" up the promotion ladder. The increasing pressure to recruit and place women, minorities, and specialists in new disciplines at all levels meant that these individuals—the ones who were successful, anyway—could not spend the traditional amount of time at each step, or the diversity sought at the higher levels would not be reached for a generation. Robertson was a district ranger in his twenties, supervisor of two national forests in his thirties, and chief at age forty-seven. Those who complained about the fast-tracking—generally white males who felt they had been unfairly passed over—would have to admit that careful selection, special training, and proper experience could and did produce a high level of technical capability and leadership skills.

SELECTION AS ASSOCIATE CHIEF

One day in 1982, Associate Chief Douglas Leisz walked into Robertson's office, saying, "I need to talk to you." He said that he was retiring and that he and Chief Peterson had "decided to send your name across the street as my replacement." Robertson was "shocked." He was only an associate deputy, and "all of a sudden I was jumping over deputy chiefs." Robertson believed that his selection was in part a reaction to the recent replacement of the head of the Soil Conservation Service, "our sister agency," with a political appointee. To ward off a similar "politicalizing" of the Forest Service, long a concern, the agency traditionally had "multiple candidates well qualified to move into any position,...to have people coming through the pipeline that are well known and supported by the secretary of Agriculture and assistant secretary." Too, the Forest Service at the time "was suffering from even-age management" at the top; the deputy chiefs were basically too old for any of them to be selected to succeed Peterson when he retired.

Robertson was well known to Peterson. "Max and I go back a long ways"—to the time when Chief McGuire had "initiated a major reorganization study of the Forest Service." The study eventually resulted in the creation of deputy regional foresters for National Forest Administration, and deputy directors for State and Private Forestry and for Research. At the time, Peterson was regional forester in Atlanta and was a member of the steering committee; Robertson was on the Mount Hood and detailed to Washington. Chet Shields, Robertson's former boss, was in charge of the project. Shields asked Robertson to be vice chairman and soon pretty much delegated him full responsibility. Peterson "took more interest in that reorganization than any of the others," and so "Max and I would get off at night and fiddle around with the organization study." They worked well together.

McGuire felt that employees ought to be involved in the study, so Robertson and others "held public meetings in all sections of the Forest Service." It was a "big deal," the agency's first public meetings with employees. "We were getting some rebellion then. There were some pretty cutting remarks about the Forest Service organization...they felt stifled and too directed and too militaristic." Good enough, thought Robertson, but "how bluntly do we put this to top management?" They decided to "reflect the tone" of the comments, and it was "probably the first time the leaders of the Forest Service got blunt feedback in an organized way." One deputy chief angrily said the study "was worthless," but McGuire was "very chiefly." He said, "We need to deal with this," and asked that the main recommendations be implemented.

The division of responsibilities between the chief and the new associate chief "kind of evolved." Basically, Robertson "just fell into Doug's mode of operation, which I was familiar with." He remembered that "Max was a traveling chief. He loved to travel." Thus, a big part of being associate chief was "being chief when the chief's out of town." Testifying to Congress was very important as well. There are many hearings that require the chief's presence, and often the associate chief would take his place. Not only did Robertson have to coordinate his testimony with the secretary and OMB, but he had to "get his own staff lined up when they're at odds with what you've got to say politically." The division of labor was not along functional lines, "it was in a constant state of flux depending on the workload, travel, and time available."

As associate chief, Robertson worked on the timber sale bailout and the California consent decree, and he examined the agency's research arm. The independence of Research had been an issue, off and on, since the division had been established in 1915. It seemed to Peterson and Robertson that Research "viewed the national forests as one of many clients." The two sat down to discuss the situation and decided that "part of our contribution to the Forest Service has got to be to get Research more responsive to the needs of our land managers." Robertson "spent a lot of time on that." Placing national forest personnel on scientists' annual performance review boards was just one method used to make Research "more responsive."

The Pilot Study Program was a "big issue" that Robertson took on. "I started this with Max's blessing but it was my idea." It was commonly accepted that "the Forest Service has absolutely the most comprehensive and complicated budget of any." At the district level, the ranger "has all of these targets and objectives...but is faced with a hundred little pots of money." Not only did Congress want assurance that certain tasks were accomplished, but also Forest Service staff officers wanted their specialties to have their "own little pot of money." A ranger once described the situation to Robertson: "I just feel like there's a big funnel sitting on top of my head, and everybody in Washington is crapping in it." The pilot study freed selected program managers to focus on getting the job done by staying within the overall budget without having to worry about the "little pots." Some deputy chiefs did not support the program, and Robertson exercised his authority as associate chief to overrule their objections. "It worked very well,...but we violated some budget laws, which got me in serious trouble." His primary objective "was changing the organizational culture," something he had felt needed doing since his experiences with politically radical classmates at American University.

SELECTION AS CHIEF

Assistant Secretary George Dunlop "was a key player." Dunlop, a "big idea guy and very innovative, creative," and Robertson had engaged in philosophical discussions, and the two "hit it off." Two years before the end of President Reagan's term, Dunlop talked to Secretary Richard Lyng and Deputy Secretary Peter Meyers about Robertson's succeeding Peterson. "They decided the time is right, 'Dale is the guy we want as our next chief,' and they just did it."

George Leonard had been director of the agency's timber management program at the time of the timber sale bailout. He came into frequent contact with Dunlop, Meyers, and Lyng as part of his responsibilities. He even accompanied the secretary to meetings at the White House; in sum, he was well known and respected by his political superiors, as of course was Robertson. "Lyng knew me and he knew George. So it was kind of a package deal" when Leonard was named associate chief to succeed Robertson as he moved up to being chief. Leonard was given more authority than any previous associate chief, except perhaps for Overton Price during Gifford Pinchot's time. In fact, Robertson recommended that Leonard be promoted to Senior Executive Service level 6, the same pay grade that Robertson held, since they both did the same job. As it happened, since Leonard had more years of service than Robertson, his salary exceeded the chief's.

In at least one area, however, Leonard did not serve as Robertson's alter ego: national security. Robertson recalled that he had "the highest secret clearance in government." The chief was one of a "small group of people" who were prepared "to go to a secret location" should a major attack by the Soviet Union take place. "We'd go through scenarios of likely hits on the United States and what kind of damage would be done with these nuclear bombs." The basic assumption was that the "president and Congress were no longer operational" and how the group would "run the country." He concluded, "These were intense exercises, very educational."

Chief Robertson and Smokey Bear. In 1987, Robertson's first year as chief, the Forest Service's Cooperative Forest Fire Prevention program instituted Smokey Bear Day with all the major league baseball teams. USDA Forest Service photo.

Robertson believed there was a myth within the agency that the chief sat at his desk and thought about the big picture and plotted strategy. "The truth of the matter, the chief doesn't have time to do any of that." Instead, "it's a hectic pace, hectic job, and everybody thinks they own the chief, and they do." Interest groups were "constantly interacting with you and demanding something, demanding your time." If dissatisfied with the chief's response, they could call their congressional representative, who would always be happy to serve a constituent and demand some more time. A single phone call from Congress or the White House could wind up consuming half a day.

"If anybody needs to be an entrepreneur, it's the chief of the Forest Service with a willingness to be fired." The chief had to be creative and innovative; "you can't be a bureaucrat" or the best you could ever be is "mediocre, and that would be on your best days." Robertson saw himself as a risk taker, and "I think most people who know me would say I took a lot of calculated risks." To have been otherwise would have made him only "caretaker" of the Forest Service. Working in the nation's capital added another dimension. "Washington prides itself on chewing up and spitting out people and demanding whatever it wants...It is a tough town."

DEALING WITH CONGRESS

Robertson believed that he took budgeting "as serious as any chief,...because big decisions get made in budget." Along with Associate Chief Leonard, "we were a force to deal with in determining what the Forest Service budget was, and the shape of it." The approval process began with sending a draft to the departmental budget officer, who would then brief the secretary and assistant secretary. The next step was to have the chief sit down with his political bosses to field whatever questions they might have. "The chief is on the firing line. You can't have a general relationship with your budget." How effectively Robertson responded to those questions "may mean millions of dollars to the Forest Service." After being cleared by OMB, the "president's budget" is sent to Congress, where the chief is the main witness in hearings before Senate and House Appropriations committees.

For a hearing, the chief and his deputies "are all lined up at the table facing the committee." Following a general statement by a political appointee, generally the assistant secretary, Robertson made his opening statement and "answered all of the questions that [I] had the knowledge to answer." When the committee looked at specific programs, such as Research or State and Private Forestry, the appropriate deputy chief would take over the testimony. During the Reagan administration, Research was controversial at budget time: Assistant Secretary Crowell forced smaller requests for that program because he believed that it did not "pay off." Members of the committee did not share Crowell's view and usually restored at least a portion of the research appropriation that the agency had originally requested.

Budgeting was "a year-long process. There's no slack season," what with a three-step sequence of authorization, budget, and finally appropriations. There was another wrinkle; ever since McArdle had been chief, the Forest Service was included in the budget for Interior and Related Agencies. "The budget makes strange bedfellows": the Forest Service had to compete not only with the National Park Service but also with the National Endowment for the Arts. While Robertson was chief, NEA fell into deep controversy over its funding of "pornography," at least pornography as determined by influential Senator Jesse Helms of North Carolina. Thus, NEA's budget was reduced, and the Forest Service budget was increased; although the chief was himself a strong supporter of the arts, he quietly cheered.

There were many other hearings throughout the year. "Congress didn't coordinate with us...and they couldn't care less about what the chief of the Forest Service's schedule is." Instead, the committee would send a letter announcing the date of a hearing and request the Forest Service to provide a witness. Robertson took each invitation to the chief and staff meeting for a decision on just who would testify. "If it was high profile, there was an expectation from the chairman that the chief would be there." The next level of importance would be handled by the associate chief. For more specific hearings about a wilderness proposal, for example, a deputy chief would be sent to the Hill. No one below the level of associate deputy chief would testify.

Partisan politics is a fact of congressional life, but most of the pressure would be on the assistant secretary or other appointee. "Although it tends to get more nasty all the time," most in Congress saw that the chief was a career professional doing a job that required testifying in support of the current administration's policy. Even so, when the administration was of one party but the congressional majority was the other, the Forest Service "gets caught up in that and it's a tough road to weave your way through." Too, in all cases the constitutional separation of powers was always in force, and deference on the part of administration witnesses was expected. Even in private, when he might be called "Dale," he never "let my guard down, and I called them Senator or Congressman or Mr. Chairman."

THE WHITE HOUSE AND OMB

"OMB is a powerful organization because they are an extension of the president, and they even order around and dictate to cabinet members." Probably OMB and the cabinet member would work it out, because "you've really got to be screwed by OMB" before resorting to a presidential appeal. Secretary Lyng and President Reagan "were big buddies from California," but even then Robertson could remember only a couple of times that the secretary took his case directly to the president.

As to the White House itself, the Forest Service maintained an ad hoc relationship. Especially early in an administration, new staffers not familiar with the chain of command would phone the chief directly. But mostly the White House talked to the secretary, who then talked to the chief. An exception was when Clayton Yeutter, President George H. W. Bush's secretary of Agriculture, was moved to the White House to be chief of staff. "Clayton talked to me fairly often" as chief of staff. If there was a Forest Service issue, "he would call me direct." That relationship had its obvious advantages but could also be a disadvantage when Edward

Chief Robertson with Secretary of Agriculture Edward Madigan, 1991. Madigan was not supportive of the agency in times of controversy. USDA Forest Service photo.

Madigan, Yeutter's successor as secretary of Agriculture, "felt threatened by it." Mostly, following the traditional chain of command was better.

In fact, "Madigan was a different kind of guy. I was dealing with a secretary like no other chief had ever dealt with." The new secretary's management style was that of "intimidation." Madigan was much bothered whenever the press carried articles critical of the Forest Service, and he felt that Robertson was doing a "terrible job of management"—otherwise such articles would not appear. Worse, in minor cases where the chief "didn't even know what [news report] he was talking about," Madigan would "accuse me of not knowing what was going on." Robertson found that all Agriculture agency heads were being treated in the same way. He would "never forget one day I walked into his office on some issue, it wasn't a big thing. Secretary Madigan just looked me in the eye and he said, 'Chief, you're nothing but a problem to me and this department. Why don't you take your organization and go to the Interior Department where you belong?'" The history of the Forest Service includes a series of efforts to move the agency to Interior, efforts that had always been resisted by the secretary. Apparently, during the final year of the Bush administration, at least the secretary of Agriculture would have supported such an initiative on that particular day.

OTHER AGENCIES AND INSTITUTIONS

Robertson "took the initiative" to establish better working relationships with the other federal agencies with related missions, such as the Bureau of Land Management, National Park Service, and Fish and Wildlife Service, all in the Department of the Interior. "Prior to the Bush administration, I would say we had interactions, periodic meetings, but it was kind of ad hoc, based on issues." Robertson was the only carryover of natural resource agency heads between the Reagan and Bush administrations, and as the new heads of the other agencies were announced, "I tried to be about their first phone call to welcome them to the administration." The four agencies, Robertson said, needed to work more closely together, and they needed to "set the pattern of cooperation for our staff" by personally interacting themselves. The three accepted.

The bureau chiefs took several field trips together, meeting regional leaders and going over issues. Each took turns hosting a trip; Robertson hosted the first gathering in Oregon on the northern spotted owl. The Park Service trip was to Alaska, BLM took the group to the Southwest, and Fish and Wildlife chose a location in Wyoming. "Even though we got crossways on some things later, at least there were some good personal relationships among the four agency heads." They had good success in coordinating activities across the boundaries between their jurisdictions.

The advocacy groups and lobbyists were good at what they did, Robertson recalled, and he generally did not feel a need to contact them: "they would usually call me." He felt that he began as chief with "very good relationships with the interest groups...but toward the end of my tenure, issues got so polarized over the spotted owl and old-growth and wilderness that I think the environmental groups decided that they had more to gain by being an adversary to the Forest Service than trying to cooperate with us." Robertson agreed that this strategy was probably a good one for them and their interests: "play hardball and be tough and don't cooperate." Under the Reagan and Bush administrations, it was the industry groups that had the influence, but that all changed with Clinton. "The Forest Service has to be pretty light on its feet."

1890s SCHOOLS

The 1862 Morrill Act donated "Public Lands to the several States and Territories which may provide Colleges for the Benefit of Agriculture and the Mechanic Arts." These land grant A&M schools would establish programs in engineering, geology, and husbandry that provided the science and practical skills so necessary to the Conservation Movement, which was then taking its first, uncertain steps.

Chief Robertson dedicating a time capsule for the centennial of the National Forest System, with Assistant Secretary of Agriculture James R. Mosely (partially obscured) looking on, Cody, Wyoming, 1991. On March 30, 1891, President Benjamin Harrison used his new authority and created what is now part of the Shoshone National Forest. USDA Forest Service photo.

Blacks were not admitted to the southern colleges, however, and in 1890 the Morrill Act was amended in a variety of ways, including creating black A&M schools that also would be funded by the sale of public lands.

Over the years the Department of Agriculture had provided research grants and in other ways supported the traditional A&M colleges, but much less so their black counterparts. During the 1960s, as equal opportunity became important, it was quickly obvious that the department was "one of the worst" in government with "mostly white males" despite the by-then long existence of all-black colleges that were graduating people with skills directly related to the department's mission.

Soon after Robertson became chief, Secretary Lyng "came out with a strong statement about how we must diversify our workforce." One of his initiatives was to recruit heavily out of the 1890s schools. The secretary convened a conference in Atlanta that included his agency heads and the presidents of the 1890s schools, plus their deans of agriculture and business administration. People from the department "made a lot of speeches" and presented a slide show that portrayed the Department of Agriculture in action. They were proud that the show was of high professional quality, but their guests saw not even one black or brown face. It was an embarrassing beginning: not only were there no minorities, but the production offered solid evidence that the department was insensitive.

Despite its shaky start, "it really was a good conference and we had a lot of interaction, got acquainted, came up with action plans." Robertson could also see that the 1890s leaders had a "woe-is-us attitude [about] how Agriculture had been neglecting them." They made their point, "except that they overplayed it." Secretary Lyng announced a task force consisting of five agency heads and five college presidents charged with developing bold programs for strengthening the schools. The secretary further stated that he would contact the original A&M schools and say that "unless you're producing the products we want in USDA, which includes minorities, we're going elsewhere to do our recruiting." Robertson "was really proud of that rather strong statement." After the conference, Deputy Secretary Meyers told the chief that he was to head the task force: "You tell me what you need from me and Secretary Lyng, and we're going to make this successful."

Not only did the task force achieve congressional authorization for direct grants to the 1890s schools, but each of the five Agriculture agencies now had appropriations that included scholarships that were "full ride." Too, a portion of summer jobs were allocated to black students "at the expense" of the traditional forestry schools. Each agency also developed a "center of excellence," selecting a college that had the "best core program." Robertson chose Alabama A&M, which had a nonaccredited forestry school but did have a Forest Service research center on campus that could hire students part-time. He thought the effort in Alabama "was very successful." Not only were they strengthening an 1890s school, but the agency was getting "the top blacks."

THE SPOTTED OWL

The report of a committee headed by Jack Ward Thomas "basically concluded that our old strategy for protecting the spotted owl was not adequate." But while considering options, "we were shut down by the judge." Robertson recommended adoption of the Thomas report, "at least initially," followed by preparation of an environmental impact statement based upon it. "I couldn't just go from a scientific report to a policy"; under the National Environmental Policy Act, it was necessary to look at alternatives and obtain public comment before deciding upon policy. Too, Congress—especially the Pacific Northwest delegation—was much interested, since the controversial report would obviously have a great impact on the region.

Robertson and Thomas both testified, but not all were convinced. Senator Slade Gorton from Washington "got really upset" that the chief was giving credence to a report by "a bunch of weird scientists," and he complained to presidential chief of staff John Sununu. Earlier, when Robertson had informed Sununu that the Forest Service "was getting into a deeper hole" over the owl, he had smiled and said to "hang in there." But Gorton's complaint somehow touched a nerve, and an angry Sununu demanded that Secretary Yeutter "fire the chief." The secretary declined but agreed to "have some discussions" about the spotted owl and options.

Robertson recalled "that it didn't help that BLM"—even though he had a good relationship with Director Cy Jamison—"took up Gorton's campaign. He came out very critical of Jack Ward Thomas and his team, saying that it was unrealistic, that they didn't care about people, and BLM was going to stand up for the people" in Oregon who would be adversely affected by a decline in timber harvest. Then Secretary of Agriculture Clayton Yeutter stepped in and took "the spotted owl decision out of my hands,...saying, 'I'll personally handle this.'" Robertson believed that the secretary "was trying to protect me because it was getting dirty and vicious." Even though Robertson did not take part in all subsequent meetings concerning the spotted owl, the secretary "always briefed me. I had a good relationship with Clayton, and he was trying to protect me."

Sununu and the White House had been involved more broadly with the owl. Logging on public lands within the range of the spotted owl had been suspended by court order. The judge stated that his order would be lifted only when the forest plans were modified to show just how the national forests would be managed to protect the species. White House factions were divided, but the Sununu-led group dominated, refusing to make changes beyond stating that such management "would not be inconsistent" with the Endangered Species Act. The judge ruled the changes inadequate and continued the ban on logging. During his campaign, Bill Clinton had promised a "timber summit" to work out a compromise. Jack Ward Thomas would continue to be closely involved.

BIODIVERSITY AND JUDGES

"In the slow ways that the Forest Service evolves over time, it would have eventually gotten to where it is now." But it was the Endangered Species Act that "really backed us into the corner." Under the Multiple Use–Sustained Yield Act, the Forest Service had recognized conflict as a part of management, but "let's come together and see what trade-offs we can make here to reach a balanced decision." With ESA, "trade-offs are not in the vocabulary"—it was "yes or no" that an "action will adversely affect" a species. "The Endangered Species Act drove an arrow through the heart of multiple use." Trade-offs were no longer an option.

The National Forest Management Act of 1976 states that national forests shall be managed in ways to "provide for diversity of plant and animal communities." Robertson believed that "if you assume that endangered species are part of your diversity," then there was a consistency between ESA and NFMA. But "laws are just a bundle of concerns that get lumped together in a piece of legislation," and legislators compromise on language as much as necessary to obtain the necessary votes. "Congress never goes through a rigorous analysis like the judges are going through now...and how it all ties together with tight logic." The only way that NFMA can be implemented "is to have this flexibility to adapt, adjust, fit, and make things work." However, "the judges don't look at it that way."

Judicial decision making meant that "one of our main clients was the judge." In the past, foresters on the ground would make management decisions, but "they weren't very good at describing these judgments." These descriptions became mandatory with the advent of NEPA, which spawned the need for interdisciplinary teams. It was a new experience, and as the Forest Service found more and more of its decisions being challenged, "we kept asking, Will this pass the judicial test?" Since "there is no such thing as a neutral procedure," the Forest Service is so "tied up" by judicial procedures that "it can no longer be efficient." Robertson believed that the American public "would be shocked to learn what it costs the Forest Service to make what really appears like a rather simple, straightforward decision." Not only were costs increasing rapidly, but since the ranger districts could not afford full disciplinary teams, more and more decisions were being made at the national forest level. Districts were no longer self-sufficient; the new requirements necessitated a degree of centralization. It was "all part of the crap that came through the funnel on top of the ranger's head."

RECREATION, FISH, AND WILDLIFE

While he was still associate chief, Robertson could see that "the two programs that had the most public support in the Forest Service were Fish and Wildlife and Outdoor Recreation." This was where the "bulk of the people are going to judge whether we're a very good outfit." It was evident to him that "we were slighting those programs,...they weren't a full partner with our timber program in terms of status." After he became chief, he announced that he would give emphasis to these programs. "That's the nice thing about being chief:...you have an idea and you have a good chance of doing something about it."

His strategy was to use partnerships, a method that would be blessed by the Reagan administration. With support from Deputy Secretary Meyers and authorization from Congress, the Forest Service inaugurated a series of challenge cost-shared grants: the agency would put up fifty percent of the cost of a recreation partnership if the amount was at least matched locally. Robertson began with ten million dollars for America's Great Outdoors Initiative, and by the end of his tenure, the federal amount had increased eightfold. There were applications for ten times the number that could be funded; "it was a wonderful idea that caught on like wildfire."

There was an internal impediment to the program, however. "The Forest Service had this hangup about we wanted concessionaires and people to come in and provide a recreation service to the public, but by God, they better not make a dime off of us." Robertson turned that around, saying that if the timber industry was "making a living off of buying trees and logging," then why not our recreation partners? It was a "major policy shift." During the 1988 presidential campaign, Robertson managed to have a copy of the Great Outdoors brochure placed in Bush's briefing book. As president, Bush embraced the program and from time to time asked how well it was doing. On behalf of the Forest Service, Robertson received the Colman Great Outdoors Award in 1989, something that he "was the most proud of." President Bush received the same award the following year.

"We did the same thing with Fish and Wildlife....We had a program called Rise to the Future, fish your national forest." Robertson was not an angler himself: "My fishing is growing up in Arkansas with a little cane pole trying to catch perch and catfish." But "the fisherman groups were so proud of what we were doing" that they contributed money. In 1988, he was named Conservationist of the Year by Trout Unlimited. "They gave the chief an award, but it's really your people out there doing a great job." But not everyone was happy with this bolstering of fish and wildlife programs: "The timber industry was always looking at me with a suspicious eye." The industry's concern was the potential of an adverse effect on the timber program. However, "I kept assuring them that we could have a better recreation and fish and wildlife program, and still carry on a timber program." Robertson believed that it was not these programs but "the Endangered Species Act that's really knocked the timber program in the head."

LAW ENFORCEMENT

"We calculated once that in the state of California the value of the marijuana grown on the national forests exceeded the value of the timber harvested in California that year. I mean, it was big business." As a result, the Forest Service "beefed up" its law enforcement division by bringing in "real professionals that were very serious about their job." It did not always work well; the law enforcement officer reported to the forest supervisor, who was used to making trade-offs. The "law enforcement folk didn't relate to that at all." By their training, the law officers viewed the visitors to the national forests with suspicion; after all, they might do something wrong. That was not the Forest Service way, and soon some officers were investigating their bosses, who apparently were not serious enough about crime prevention. But there, in fact, was some wrongdoing by a few forest supervisors themselves, and that news "got to Congress."

Robertson found his next appropriation encumbered with a rider that required him to establish law enforcement as a separate organization, in which officers reported to other officers instead of to forest supervisors. This was the "stovepipe" model, where authority went directly from top to bottom, instead of the four administrative levels common to all other programs. "My forest supervisors thought I sold them down the drain on that, but I didn't." Instead, he had lost the battle with Congress, an uneven struggle at best. After the officers became separate, "they kind of took on a life of their own." Chief Thomas would also have problems with the agency's law enforcement program.

WILDERNESS PURITY

"Wilderness has been a blind spot in the Forest Service." Even though the Forest Service had developed its rationale and philosophy, "somewhere along the line wilderness became a threat to multiple use." Robertson well remembered that when he was a young forester in Oregon, his first regional forester said that "wilderness was infringing on our multiple-use management prerogatives." The chief believed that the "purity issue"—only the most pristine of areas could qualify as wilderness—had been raised as a "defense mechanism" to keep wilderness acreage at a minimum.

When Robertson had been associate chief, he became involved with the Environmental Protection Agency's request to use helicopters to sample water in wilderness areas. "Max was taking a hard line." When Robertson met with the EPA administrator, he was asked, "Don't you have any flexibility?" After a detailed examination of the proposed sample sites, they had "narrowed it down to a very few lakes, and I finally agreed to let them" take the samples. When he reported to Chief Peterson on his decision, "Max was relieved to get out of the situation by that time, because it was in the papers and we were being painted as inflexible."

As chief, Robertson found himself being "beat up" when he went to Congress because the Forest Service had "carried forward with the purity concept of wilderness to the point of being ridiculous." He would hear from members of committees that the agency "was not being realistic." A case in point was that outfitters and guides were big business in wilderness, but under agency regulations whatever was packed in had to be packed out. They asked Robertson to "change the policy to be more reasonable" and give them permission to cache basic camping materials, such as tents and sleeping bags, over winter.

"I worked with the folks and, man, I was fighting a one-man battle." However, he was able to get the policy changed to allow overwinter caches. Some Forest Service retirees who had been active when the Wilderness Act was passed in 1964 "got on a campaign" that "the chief sold us down the drain," that Robertson "had no respect for wilderness."

At times, even the most vocal of wilderness advocates disagreed with the Forest Service on the purity issue. A ceremony was scheduled to name a peak in the Sierra in honor of an elderly and by now handicapped woman who had climbed it many times. Not only were several wilderness groups involved, but a member of the California congressional delegation would be present. But the peak was in a wilderness area, and the Forest Service denied a permit request for helicopter transport, now necessary to bring in the woman, "the star of the show." Robertson was told that the Wilderness Act forbade the use of helicopters; he examined the text himself, and pointed out that in fact the law allowed a degree of flexibility for "administrative purposes." To the chief, the ceremony was an administrative purpose; "there was nobody fighting it except the Forest Service people and this mentality they walked around with." Over much internal objection and little support, he "widened the flexibility to deal realistically with some things in wilderness management."

TRADITIONAL FORESTRY HITS THE WALL

"Traditional forestry would no longer fly in the federal government," Robertson observed. Clearcutting continued to be highly controversial, "no matter how much we foresters thought that it was good scientific forestry." To the American

Chief Robertson and Alexander Isayev, Soviet minister of forestry, displaying the forest type map of the United States. USDA Forest Service photo.

people, "it looked like abuse of the land." The Endangered Species Act was the "real driver" because of the "hammer that it had." Environmental groups had "it all figured out, and they used the hammer through the courts." All the technical debates about endangered species aside, Robertson came to the conclusion "that the multiple-use management that the Forest Service was practicing was creating endangered species." Too, there was the National Forest Management Act, which mandated the maintenance of viable populations. Robertson's declaration that clearcutting would no longer be a standard practice on national forests was not adequate; forestry had "hit the wall," and incremental shifts of policy and practice were too little too late.

With passage of NFMA, Congress thought it "had put an end to clearcutting, at least in a massive way. But then the Forest Service and the timber industry got in there piddling around with the wording." Clearcutting was to have been an exception, and the "Forest Service took maximum advantage of that exception," which was a "misjudgment." The Forest Service had a choice with NFMA. It "was a major juncture," but the agency decided that eliminating clearcutting was not necessary.

"We had to have a new concept." Scientist Jerry Franklin had been recommending adoption in the Pacific Northwest of what he called New Forestry. Robertson "grabbed onto New Forestry because it was the only new concept "that was on the table." Without full consensus within the agency, New Forestry evolved into New Perspectives. Robertson did not particularly like the term, but he was testifying in Congress the next day, and he and his staff had not come up with a better label. When he testified about what the Forest Service was doing to address the endangered species issue, he reported that New Perspectives provided a new and broader way of looking at and managing the national forests.

New Perspectives was "kind of a pilot test." On a case by case basis, scientists got together with land managers to "test some alternative ways of managing the forest and harvesting timber at the same time to provide for these other values in the broadest sense." Robertson was favorably impressed by what he saw, and New Perspectives pilot tests continued for several years. The next policy evolution grew out of a presidential announcement at the U.N. Conference on Environment and Development.

In June 1992, EPA Administrator William Reilly led the U.S. delegation to the "Earth Summit" in Rio de Janerio. He was "as much of an environmentalist as the Bush administration had," and he and Robertson had previously formed a friendship, occasionally meeting at the White House for lunch to chat about issues of mutual interest. Reilly would often question Robertson about clearcutting, but the chief admitted that his "explanations weren't all that convincing." He told Reilly that he felt that clearcutting, as well as endangered species, could be dealt with through the much broader framework of New Perspectives, which by then the Forest Service was calling Ecosystem Management.

Senator Al Gore attended the Rio summit, too. He "lambasted the United States about what terrible forestry practices we had in this country," aiming to embarrass the Bush administration "in front of the world." President Bush was to speak during the last day, and Reilly was explaining by phone to Chief of Staff Yeutter that the president needed to defuse the criticism and "say great things." Yeutter then asked Robertson for a statement "to eliminate clearcutting that the president can announce in Rio. Boy, the lights went on." Robertson well understood that new policies had to be formulated but could not be adopted and implemented without "working the process" in Washington, which included his political bosses and Congress. "There was my chance to get the official policy." Not only would the president announce the end of clearcutting as a standard practice on national forests, with some exceptions, but he would also explain that it was all part of a new policy called Ecosystem Management. He drafted the message for Yeutter, and the president agreed to the language. Chief Robertson made the announcement on adoption of Ecosystem Management at home, and the president confirmed it in Rio. "That was really something, to get it through the Republican administration."

There had been a glitch, however. Secretary of Agriculture Madigan "was not in the loop." It had all happened so rapidly that there had been no opportunity to brief the secretary. Before making his public announcement on Ecosystem Management, Robertson first went to Assistant Secretary John Beuter, who agreed to the "need for a major change" in policy. Beuter and Robertson then went with the secretary's public relations officer to explain the situation to a secretarial aide, who exclaimed, "What's this ecosystem stuff!" He hoped that there would not be "any words like that in the press." The public relations officer assured the aide that "clearcutting" would be the headline. The next day, the *Washington Post* headlined "Ecosystem," relegating clearcutting to a subtitle.

TRANSITION FROM BUSH TO CLINTON

"The night that Clinton announced Al Gore as his running mate for vice president, I knew I was gone if they were elected." Chief Robertson had had "run-ins" with Senator Gore, who was leader of a group of environmentally oriented senators. Each year the group had tried to have the Forest Service appropriation for timber and roads "drastically reduced." Robertson had to "call every friend" in the Senate to eke out narrow victories over "Al Gore and his band of environmentalists." He believed that there was "bad blood between me and Gore and that group of senators." He added, "I wouldn't have been an effective chief if I had let Gore just run over the Forest Service."

The handwriting was clearly on the wall following President Clinton's inauguration. During the campaign, he had promised a "timber summit" to find a solution to the spotted owl impasse in the Pacific Northwest. Chief Robertson attended the high-profile event, but only as a member of the audience. At the table was Secretary of Agriculture Michael Espy, who had been briefed on the issue by someone other than the chief. Later the secretary testified to a congressional committee that "he was going to have to change the leadership of the Forest Service," but he had not discussed the matter with Robertson ahead of time.

It fell to Assistant Secretary James Lyons to confirm the rumors, face-to-face, with Robertson. In Robertson's assessment, "they didn't know how to handle it, and it got sloppy and personal." Robertson and Associate Chief Leonard were both assigned to "other duties" and soon retired. Later, Secretary Espy resigned following charges that he had accepted illegal gifts. He was indicted but found not guilty.

Only fifty-four when his Forest Service career ended, Robertson explored his options. He decided against formal employment, choosing instead to "just enjoy life" in Sedona, Arizona. His only regular "job" is as a volunteer member of a Forest Service trail maintenance crew. He is especially delighted that the crew boss is a retired National Park Service employee.

CHIEF JACK WARD THOMAS

USDA FOREST SERVICE PHOTO

Jack Ward Thomas will no doubt always be linked to the northern spotted owl. This reclusive bird accounted for the rise to prominence of a wildlife biologist best known previously for his work on elk, and the controversies that ensued from his study of the owl and its old-growth habitat followed him into the chief's office.

Thomas was born in Fort Worth, Texas, on September 7, 1934. He graduated from Texas A&M in 1957 with a bachelor of science in wildlife management. Upon graduation, he went to work for the Texas Parks and Wildlife Department, where he stayed for ten years. In 1966, he began his Forest Service career in Morgantown, West Virginia, as research wildlife biologist, and at the same time earned a master's degree in wildlife ecology at West Virginia University. Four years later, Thomas moved to Amherst, Massachusetts, as principal research wildlife biologist, and entered a Ph.D. program in land-use planning at the University of Massachusetts. A 1973 article that he coauthored with Robert Brush and Richard DeGraaf, "Invite Wildlife to Your Backyard," was published in *National Wildlife*, and it became the most reprinted article in that journal's history.

Thomas's next transfer was his final one as a Forest Service scientist. In 1974, he moved to La Grande, Oregon, as chief research wildlife biologist. The experiment

....

Source: An Interview with Jack Ward Thomas, *by Harold K. Steen. Durham, North Carolina: Forest History Society, 2002. Thomas was interviewed in 2001.*

station director asked Thomas to bolster the wildlife portion of the existing wildlife and range research program, which until then had pretty much emphasized range issues. His prominence as a scientist continued to grow, and in 1989 he was invited to head an interagency team to study the spotted owl, already an issue of great controversy. Four years later he would succeed Dale Robertson as chief.

Thomas keeps a journal of his activities and thoughts, to date filling more than three thousand pages that recount his professional and personal life.[1] A frequent topic before he became chief—less so afterward—is hunting, especially elk. These superbly written essays capture his thoughts about nature and wilderness, at times while sitting with his back to a pine and warmed by the sun, with rifle across his lap. It often seems that the outing is more important than the hunt, just getting away with good friends and thinking about more valuable things than work.

SELECTION AS CHIEF

George H. W. Bush was president in 1989 when Thomas began to study the spotted owl. A court order had suspended logging on public lands within the range of the owl, bringing enormous pressure on Chief Robertson to "fix" the problem, mainly the real threat to the timber economy of the Pacific Northwest. Although the chief survived an effort by the Bush White House to have him ousted, when Bill Clinton succeeded Bush in 1993, Robertson and Associate Chief Leonard were soon removed. During his interview Thomas commented on the manner in which Robertson was forced out. "It was clumsy. It was demeaning. I resented it then, and I resent it now."

By this time the Jack Ward Thomas report, as the study by the Interagency Scientific Committee to Address the Conservation of the Northern Spotted Owl was generally called, had provided Thomas with a renown far beyond the world of science. Too, there was FEMAT—the Forest Ecosystem Management Assessment Team—which Thomas also headed at the request of James Lyons, who was staffing the Clinton timber summit in Portland. FEMAT prepared a range of options for balancing the owl with the economy, and the Clinton administration selected Option 9. By then Lyons was assistant secretary of Agriculture with responsibility for the Forest Service and was urging a reluctant Thomas to become Robertson's successor. In December 1993, Jack Ward Thomas became chief of the Forest Service.

---

[1] *See* Jack Ward Thomas: The Journals of a Forest Service Chief, *edited by Harold K. Steen (Forest History Society and University of Washington Press, 2004).*

Thomas was not a member of the Senior Executive Service, the leadership cadre of federal agencies since 1979. For that reason, he was not eligible to be chief via the normal appointment process; he needed to be a Schedule C, or presidential, appointee. Even before becoming chief, Thomas strongly objected to being a political appointee, and he was promised membership in the Senior Executive Service in due course. As it turned out, poorly understood technicalities in the regulations on SES eligibility made it impossible for Thomas's appointment to be changed, despite support from at least as high as the office of the vice president. This situation, plus the frustrations of being chief, prompted him to submit his resignation after two years. Secretary Daniel Glickman persuaded him to stay another year, until after the November 1996 election, by threatening to replace him with an "outsider."

THE CHIEF'S DAY

"Chiefs are at the beck and call of people in the political hierarchy above their level." Sometimes Thomas would be diverted from his anticipated schedule "to run over to see Lyons" or perhaps the secretary himself. Or there might be a suddenly called meeting at the Council on Environmental Quality or a need to go up to the Hill. The "typical day simply didn't exist. It was something akin to controlled chaos." There was another factor that contributed to confused schedules: although Forest Service employees generally worked from seven or eight in the morning to five or six in the evening, the politicos "don't show up until midmorning or later, and then they work until eight at night or later."

Although they often lacked managerial experience, Thomas thought that the political appointees "worked long and hard," but he called their actions at times "lurch management." Agency employees "tried to proceed methodically to carry through some program" but had to respond to "the politicians whose power and presence was transitory, and it was somewhat difficult. But that's our system." Another part of the system was that the chiefs, though selected by political appointees, did not see themselves as political appointees with responsibility to carry out an administration's "agenda." Thomas was once told, "We don't think you're one of us." Well, Thomas didn't think so, either; in fact, he saw himself as a chief who was to guide the Forest Service for the long run, not just until after the next election.

DEALING WITH POLITICAL BOSSES

"The Forest Service in general thinks that the Forest Service is a 'department'—and that is sometimes a problem." Not only does this sense of independence

Elk hunter Thomas with his horses, Monita and Summer, 1979. Hunting was Thomas's favorite escape from daily pressures. Photo courtesy of Jack Ward Thomas.

frustrate secretaries of Agriculture, it drives other departmental employees "nuts when Forest Service employees say they work for the U.S. Forest Service. The politically correct title is USDA Forest Service. It's a bone of contention. Maybe everybody ought to relax."

While chief, Thomas reported to Secretary Michael Espy and then Secretary Glickman. Espy was "aloof," and the chief saw him only twice, during an awards ceremony and during a meeting with the Alaska congressional delegation. Glickman replaced Espy, who had resigned under fire. Suddenly, Thomas saw "a whole lot more of the brass in USDA," including the secretary and his staff. He could see the secretary on short notice if he asked to see him on a priority basis. However, usually he would work through Lyons, trusting the assistant secretary's judgment whether the secretary himself needed to be involved in a decision. But if he had it to do over, he "would have seen more of the secretary. Access is power."

Initially, Thomas did not meet with Secretary Glickman on a regular basis; rather, he would request a meeting only if there was a pressing issue. However, the chief was invited to Agriculture staff meetings, but he "would rather eat worms than sit through a general staff meeting in USDA." Even though half of the department's employees worked for the Forest Service, "ninety-five percent of the conversation would concern farm programs, food stamps, and other such topics. There was almost never any discussion of Forest Service matters." Eventually, Glickman agreed to meet with Thomas on a regular basis.

"Activist undersecretaries make significant decisions," Thomas remembered. During the Reagan administration, Assistant Secretary John Crowell "caused the Forest Service to place lands in the timber base that were not appropriately productive, and squeezed the system." It was all part of "the doctrinaire belief from political overseers who said we could do twenty-five billion board feet per year." Although the amount actually cut was much less than Crowell's goal, it was "too high to be publicly acceptable to the environmental side and too low to satisfy the timber industry." It was the Forest Service's "Vietnam." It fell to Chief Robertson, and then Thomas, to pick up the pieces.

Some political appointees were "exasperatingly young," and some had little education or experience that would help them understand the Forest Service mission. Assistant Secretary Brian Burke was a case in point. Even though he and Thomas got along well, Burke had "no background in natural resources—none, zero, nada." Thomas thought that Vice President Gore had sent him from his own staff "to help bring the Forest Service into line." Thomas agreed that he had never "completely understood" just what the administration wanted the Forest Service to do because of the "conflicting signals" it was sending out. "There were times when I didn't think that the right hand knew what the left hand was doing."

Two political appointees—the secretary of the Interior and the director of the Council on Environmental Quality—traditionally had little direct contact with the chief of the Forest Service. That changed with the Clinton administration. "Oddly, I probably met more with Secretary of the Interior Bruce Babbitt than I did with Glickman, for varying reasons." One of the primary reasons was the weekly meeting convened by CEQ Director Katie McGinty—dubbed by the Forest Service as the Tuesday Afternoon Club—at which Thomas would meet with an ever-changing mix of administration leaders with interest in the environment. He was very impatient with these meetings, which lasted for hours, because "within twenty minutes everybody had said everything they had to say that amounted to a pinch of salt. Then we would sit there and continue to plow all around it for hours more." Too, Thomas received orders from both Babbitt and McGinty, and he was uncertain whether Secretary Glickman had deferred to them, but it was clear that they had the necessary authority. However, both Babbitt and McGinty treated the chief with respect, even though they might disagree on certain issues.

President Bill Clinton with Margaret and Jack Ward Thomas, Oval Office, 1993. White House photo.

Thomas also had an unusually active relationship with Mike Dombeck, acting director of the Bureau of Land Management. The agencies had common interests—the spotted owl, Columbia River salmon, and forest fires, for example—but such overlaps had always existed, and historically, the two agencies had seldom done more than keep each other informed. Thomas and Dombeck were "probably the closest of any chief and a BLM director." An important link was Robert Nelson, director of Fish and Wildlife for the Forest Service. Dombeck had worked for Nelson before he moved to BLM. Thomas believes that Nelson was "probably instrumental" in Chief Robertson's inviting him to head the spotted owl committee. Too, Nelson invited Thomas and Dombeck to join him on hunting trips; the three wildlife biologists worked and played well together. Dombeck would follow Thomas as chief.

Friendship and cooperative spirit aside, the administration seemed to view the Forest Service and BLM through different prisms. For example, timber salvage was extremely controversial for the Forest Service but much less so for BLM.

"The administration chose to scapegoat the Forest Service and let BLM off the hook," Thomas believed. He was required to attend the weekly CEQ meetings to report on how he was meeting salvage targets, but "Dombeck was seldom, if ever, present." The Forest Service was "left hanging out to dry," but BLM "escaped the hits."

Thomas called the Department of Justice "a political arm of the government." DOJ attorneys alone represent federal agencies in court; departmental attorneys act only in an advisory capacity. DOJ exercised "great political clout" by deciding how vigorous a defense to make. Thomas could remember lawsuits that the administration wanted the Forest Service to lose, and in that indirect way institute policies that were compatible with those of the president. A case in point was a suit brought by Friends of Animals to require the Forest Service to prepare an environmental impact statement "on bear baiting as a hunting technique."

To Thomas, the suit was really a challenge to the long-standing division of responsibilities between the states and the federal government by which the states managed nonmigratory game and the federal agencies managed the habitat. In response to the lawsuit, DOJ assigned two young and, to Thomas, naive litigators, who announced without talking to the chief that "they were not going to defend the Forest Service in this case." Then, when the Forest Service lost, it would be compelled to prepare an impact statement that would reveal shooting baited bears "to be abhorrent." Thomas was able to convince the lawyers that the case was really a ploy to get a federal foot in the hunting door, an important first step in federal control of hunting of all kinds.

The lawyers checked with their bosses and reported back to Thomas that they would defend the Forest Service after all. To bolster their case, Thomas talked with former Chief Peterson, who was executive director of the International Association of Fish and Game Agencies. Peterson arranged for an intervener, who "did a fine job and won the case," while the DOJ attorneys "performed as expected." Thomas wished that the citizenry understood "the political power vested in the DOJ in their ability to defend or not to defend or how vigorously."

During the first days he was chief, Thomas sent a message to all agency personnel "to obey the law and tell the truth." Suing the agency "had become quite common, and we were losing much of the time." By definition, Thomas said, when you are sued and lose, you "didn't obey the law." The Forest Service was being told "time after time" that "it was not in compliance with the law." In those cases that were lost because of procedural deficiencies, an obvious remedy would be to ensure that procedures required by law were followed more carefully. "After you get busted several times for the same procedural error, an intelligent organization or person would say, We're not going to make that mistake again—that is, 'obey the law.'" Too, the agency should be more forthcoming when errors have been made and "tell the truth."

DEALING WITH CONGRESS

"Personally, I relished testimony. I loved it." Testimony seemed much like a doctoral exam: both required thorough preparation, and his staff provided detailed information for his study. "It was a game, a sophisticated game at which I excelled, if I do say so myself." When the Republicans took control of both houses, Democrats generally lost interest in their committee assignments, and so Thomas found himself testifying to Republicans "who were severely agitated by declining timber harvests." He recalled that Alaskan Donald Young, chairman of the House Committee on Energy and the Environment, "came after me like a tiger." Young opened the hearing by warning Thomas, "Chief, I've thirteen mad Republicans and no Democrats. I think that this is going to be a very long day for you." Thomas responded, "Mr. Chairman, as I count there are fourteen, counting you. That makes the odds about even. Let's get on with it." Thomas believed that congressional committees generally enjoyed his forthrightness, knowing that he would not try to "slip the questions" in the way that the politically more skillful Lyons could do. Thomas thought that his responsibility lay with providing full answers, even if they might not fully support the administration's position on an issue.

Thomas "kind of flinched" when testifying with Lyons "because there were people on those committees who intensely disliked him," which could poison the hearing. However, the assistant secretary—Lyons would soon be promoted to undersecretary as part of an overall departmental reorganization—"had some IQ points on several of the attackers and could come out on top of those exchanges." Eventually, congressional anger reached the point of "defunding" Lyons; officially, he could no longer receive a salary for carrying out his Forest Service responsibilities. "It was a political joke, but it was a marked expression of displeasure by the opposition party. I think that Mr. Lyons considered it as a badge of honor."

REINVENTION OF GOVERNMENT

Thomas believed that the high-profile "reinvention of government" program, led by Vice President Al Gore and intended to make the federal bureaucracy work better and cost less, "had almost no good result as far as the Forest Service was concerned." For example, "forced targets for reduction in personnel" included "big buyouts." More significantly, "the numbers are right back where they were." The biggest hurdle was Congress, which continued to ask the Forest Service for more while insisting that there be fewer employees, especially in the Washington office, which was seen as "overhead." And then there were the congressional queries; a letter that took twenty minutes to write might "cost the Forest Service ten man-months to answer." Thomas thought it would be good if "Congress would also be held accountable for costs associated with such requests": the Forest Service, he said, should include the cost of preparation in the report itself.

Reinvention "was something of a bust. It kept us distracted from our assigned base tasks for a year." Reinvention activities were disruptive because "we kept heaving a hand grenade labeled 'reinvention' into this unit and that unit."

THE SPOTTED OWL

In 1989, when Thomas was asked to head the Interagency Scientific Committee to study the spotted owl, he "hesitated to take the assignment." He had just launched his "dream study" at La Grande, and he did not want to be away for the six months or so that ISC would require. Too, he knew that his "life would never be the same, as I suspected the required alteration in forest management would be dramatic." Job loss could be high in timber-dependent communities, and he might become "notorious." He could already hear the chants, "Hey, hey, JWT, how many jobs have you cost today?" He also knew that "fading back into obscurity" in La Grande afterward was doubtful. "This was crossing the frontier between science and management."

He believed that the initial assignment, to deal with only the spotted owl, was flawed. The committee disagreed with this "narrow mission." The scientists could see that the question was not owls, it was "the old-growth ecosystem." To study only the owl would yield a single solution instead of an array of alternatives. But the land management agencies were "up against the wall," and the Forest Service had "lost its credibility" because of political and legal gridlock. Thus, the request for a single answer.

Federal Judge William Dwyer in Seattle knew better as well that concentration on a single species "was not responsive to the real question of the myriad species reliant on the old-growth ecosystem." The judge was familiar with the Endangered Species Act and knew its purpose—"the preservation of ecosystems upon which threatened or endangered species depend." The next step was creation of the Forest Ecosystem Management Assessment Team, which was to deal only with public lands; that is, the problem was to be solved with public lands alone. The Fish and Wildlife Service, whose scientists had been directly involved

The Interagency Scientific Committee to Address the Conservation of the Northern Spotted Owl in the chief's conference room, Washington, D.C., 1990. From left: Eric D. Forsman, E. Charles Meslow, Barry R. Noon, Jared Verner, Jack Ward Thomas, and Joseph B. Lint. Thomas's spotted owl assignments led to his being named chief. USDA Forest Service photo.

in the various studies, decided that state and private lands needed to be included. "It was a steep learning curve. Our instructions tried to confine the damage to timber yields and public lands. By the time we were through, we had arrived at ecosystem management. That sequence of events changed the way natural resource managers viewed the world. I don't believe there is any going back." Chief Robertson had been caught in a paradigm shift. "He was trying to take us there. Unfortunately, his string ran out before he quite got the agency there."

LEOPOLD'S LAND ETHIC

In 1995, the agency published *The Forest Service Ethics and Course to the Future.* Thomas was chief at the time and believes that the book "certainly reflected the Leopoldian land ethic." Thomas thought that Aldo Leopold had extended the notion of ethics beyond its traditional human focus "to other things, to animals, or collectively to the whole organism—the land." Ironically, while a student in wildlife management at Texas A&M, Thomas "had serious reservations" about Leopold's classic, *Sand County Almanac*. He and his classmates considered it "maudlin. We thought it was a bit weird and strange." The students saw themselves as "hard-core young scientists being trained to go forth and fulfill our role in making the world a better place in terms of the application of science and technology to wildlife management." Years later, Thomas reread Leopold's words and "saw things that I had not seen before." Since then he has reread *Sand County Almanac* every year on his birthday.

To Thomas, the evolution of Leopold's thinking is made even more "fascinating" because "he reached maturity in the Forest Service." Leopold, a Yale Forest School graduate, had worked on southwestern national forests before moving to the Forest Products Laboratory as assistant director. Then he changed jobs but stayed in Madison, shifting from the lab to the University of Wisconsin, where he taught game management. Thomas doubted that Leopold had spent much classroom time lecturing every day on ethics itself. Instead, he was teaching "hardcore botany and forestry." Thomas believed that "the environmentalists who have made him a deified character would probably be shocked and not at all pleased that he was brought to these interests and insights through his life as a hunter."

Thomas, too, had evolved since his college days. During his first job with Texas Parks and Wildlife, he had practiced a policy of predator control. In fact, golden eagle talons hung from the rearview mirror in his pickup. When he "rediscovered" *Sand County*, he removed the talons and buried them.

ROADLESS AND WILDERNESS AREAS

"The Forest Service was essentially through entering roadless areas in 1993." The agency had lost its road budget and got only a portion of it back, by just one vote in the House. "We were digging ourselves into a deeper and deeper hole on deficit timber sales and associated roading." The political, economic, and biological facts of life were part of the roadless policy; there was no use maintaining the fiction that forests in remote areas ought to be carried in the timber base, so most roadless areas would remain uncut with or without a new policy.

But Thomas believed these remote areas needed a better name—backcountry or watershed, perhaps—and their management needed to be articulated beyond banning roads. After all, the areas came without roads, and management ought to focus on what was done, instead of what was not done. He saw the issue as zoning, which had altered the meaning of multiple use. "We zone wilderness. We zone roadless areas. We zone national recreation areas. We zone national monuments.

We zone wild and scenic rivers. The only thing that we haven't zoned is the timberland." As to the roadless area controversy specifically, it "has been a huge political game since its very beginning. That game is over for the time being." The headlines that suggest roadless areas are being threatened was just part of that game, when in fact the "backcountry people have what they want."

One type of backcountry—born of controversy and still contentious—is wilderness. Like roadless areas, wilderness areas need to be managed. When he became chief, one of Thomas's innovations was to make the Frank Church River of No Return Wilderness a single management unit, instead of being managed "as a sideline by several national forests. It would be a core management unit in and of itself." But the new chief did not understand the need "to kiss all the right rings." The guides' association urged Senator Larry Craig from Idaho (the wilderness area had been named in memory of Idaho Senator Frank Church, whose guidelines on clearcutting had provided language for the National Forest Management Act) to insert a line in the Forest Service budget that said "no money could be used to execute that decision." Thomas thought that it was "a dumb thing for the guides to do. But it was my fault, as I should have talked to them first."

Thomas agreed with Chief Robertson that at times the Forest Service had been inflexible on wilderness. He remembered that Chief Peterson had constrained the Environmental Protection Agency's helicopter access for water sampling. Thomas himself had "overruled the regional forester in the Southwest when he prepared to issue a special permit to allow bulldozers to dig water tanks in the Gila Wilderness. "There's a tight line that you walk in dealing with wilderness," and sometimes "we hew too tightly to that line."

Thomas recalled an incident in which a Boy Scout became lost in a wilderness area, and the Forest Service authorized a helicopter search. When the pilot spotted the lost Scout, he radioed for permission to land. After telling the dispatcher that the Scout appeared to be in good condition, the pilot was instructed to drop him a note saying that a trail crew would rescue him the next day; permission to land was denied. The next day, the crew was unable to locate the Scout, and another helicopter search was authorized. By now Chief Thomas at his desk in Washington, D.C., was in the loop; in fact, the media were portraying the event as the "government idiocy *du jour.*" He thought that the pilot was in effect an incident commander, who like fire bosses should be able to make a "decision on the spot to pick up that kid or not." It became necessary for him to overrule strong staff opposition to helicopter landings. He could see that "some of these absolute purity arguments are quasi-religious in nature. We're trying to apply pragmatic, practical sense to a 'religious' discussion, and it simply doesn't fit very well."

WORKFORCE DIVERSITY

Thomas "went to Washington to work on resolution of the great natural resource issues of our time." However, shortly after becoming chief he was told that "our primary mission was civil rights and meeting the secretary of Agriculture's objectives in that arena." Although he tried his best "to carry out the civil rights agenda," he thought that "it was going too far" to place that goal ahead of all others.

"It was a difficult situation." The National Environmental Policy Act necessitated hiring staff from an array of disciplines "at the same time pressures were building related to workforce diversity." However, as the Forest Service began hiring social scientists, landscape architects, geologists, fisheries ecologists, and on down the list of "ologists," it found that "women and minorities were more common in these professions than in forestry and engineering." Overall, the Forest Service "has done very well with women but has not done nearly so well with minority groups." Too, the agency has tried "to meet a social objective placed on top of a system that should be blind to all but hiring the best-qualified people." In theory, the Forest Service would not place race ahead of qualifications when hiring, but "anybody that won't admit that we 'stretched' it a bit to meet racial and gender hiring goals wouldn't be quite honest."

"Civil Rights Mafia" was a term that Thomas said some people used "under their breath," as it certainly would not be "politically correct" to use the term openly. "There is an entire bureaucracy within the government that deals with civil rights. They are sort of the civil rights enforcers." The group made the "body counts" on promotions as well as who failed to be promoted, and why. "They determined the level of civil rights complaints and decided how complaints were settled and that sort of thing."

Thomas had found out "very quickly" about political correctness. He liked to tell jokes, but he learned, painfully, to "never tell jokes—never." The turning point came immediately after he joked while announcing an appointment. Not only was the individual well qualified by experience, Thomas said, he "liked the way he cut his hair." There was general laughter from the audience—both the chief and the appointee were balding—but "three civil rights complaints were quickly filed in reaction to that comment." The women who filed the complaints argued that since only men become bald, Thomas had selected this person because he was a man. It was indeed "a difficult situation."

SECURITY AND LAW ENFORCEMENT

Even before he became chief, Thomas received threats on his life. "After the spotted owl report was issued, the FBI advised me not even to go back to my home for several weeks." With his wife and children, Thomas was sequestered in a resort. Another time, following telephoned threats, one morning he noticed that the "hood on my pickup was cracked open. I looked and could see three sticks of dynamite." As it turned out, the explosives were a bundle of road flares, "but the point was well made."

The threats continued. As chief, "usually I traveled by myself, but sometimes I had a bodyguard with me. I left those decisions to others." Thomas wanted to ignore the threats, but Forest Service law enforcement agents insisted that at least a few of them could be authentic. Whether he traveled with an escort depended upon "where we were going and what the agents were picking up out of the intelligence nets." The agents served in other roles, "making sure transportation was squared away and that I was going to get to where I needed to go in the most efficient and safe manner."

The Law Enforcement division has much broader responsibilities than protecting the chief. Thomas believed that the agency did not have "any alternative to a skilled and effective law enforcement branch." The choice was between having law enforcement professionals, or Forest Service people "playing a role in law enforcement." During his research days, if he spotted someone doing something wrong, he "took care of it," even writing tickets. Times change, and now "you would be liable to get killed at worst and sued at least." Crimes that used to be limited to city streets were happening on national forests. "So that's why we have law enforcement officers."

Timber theft, not a new issue by any means, took on a vexing twist while Thomas was chief. Earlier, Robertson had established the Timber Theft Investigations Branch, which worked in the Pacific Northwest on some "high-profile" cases. "Then personnel problems began to surface." Thomas met with TTIB representatives. The director of Law Enforcement had recommended moving TTIB back into his division. Thomas consulted with the Office of Inspector General, who also recommended disbandment. Thomas accepted the advice, but some members of TTIB "went public and said we were covering up timber theft." The accusations made the press, but "we were unable to respond because of the 'rules of the personnel game.'" The "flap" played out, and when Thomas retired, employees of the Law Enforcement and Investigations division chipped in to purchase a silver-plated .357 magnum as a gift to show their gratitude for his broad support of the group.

NATIONAL GRASSLANDS AND GRAZING

The national grasslands, originally acquired by the Soil Conservation Service during the Dust Bowl days of the 1930s, were in 1960 transferred to the Forest Service, which had long experience in managing range. The grasslands were "sort of a stepchild" because the agency apparently lacked specific management authority. "So in lieu of any other direction, the Forest Service decided that we would manage the grasslands under the management authorities given to us in our legislation." Some grazing permittees in the Dakotas, having gained support from their congressional delegations, challenged managing the grasslands "under the same laws and under the same scrutiny" as the national forests. North Dakota Senator Bryan Dorgan and others "became rather vociferous that the Forest Service was going to turn over more management authority to these permittees. We were not willing to do that."

The broader issue of grazing on the national forests included a similar thrust to "somehow put more ownership 'rights' in the hands of the permittees grazing federal lands. That idea keeps coming back like a bad penny." The situation became serious enough that Thomas and BLM Acting Director Dombeck spent a "whole weekend on our collective response." Their effort was successful: "I would claim that Dombeck and I personally thwarted that end run on the public's lands. And we happened to make some fairly hefty enemies in the process."

ENDANGERED SPECIES AND COLUMBIA SALMON

The Endangered Species Act played a significant role in turning Thomas's career path from research to management. He thought that ESA could be improved by altering the definition of a recovery plan to include both long- and short-term risk. "Management agencies usually want to take a little more leeway, a little more risk for a better result over a longer period of time. The regulatory agency wants to take little or no risk over a very short period of time." He thought that a long-term risk analysis would produce "a very different answer, at least in some cases." The technical point was that "ecosystems are indeed volatile and are changing," and thus certain short-range management actions may well be inappropriate for the long run. But the act allowed very little flexibility.

Thomas also believed that ESA needed "some sort of appeals body," that the current so-called God Squad—mostly composed of cabinet members—was not practical in that it was not able to spend the necessary time to assess an issue. "Ultimately, when you lay down the recovery plan and designate critical habitat for one endangered species over another endangered species, chaos finally ensues." Much needed was recognition of the fundamental importance of endangered habitats and ecosystems, a broader issue than a particular endangered species. When the Forest Service adopted ecosystem management during the Robertson administration, it placed forest management within a much broader context that went beyond traditional policies. Like the replaced Forest Service policies, at times ESA was too specific and too narrow.

When the spotted owl situation in the Pacific Northwest had been resolved, for a time at least, the Clinton administration shifted to the "unfinished environmental issues" of the East Side, the eastern portions of Washington and Oregon. PACFISH was an earlier, interim program that directed "how to deal with timber activities and grazing issues" wherever salmon habitat "was of interest." PACFISH encompassed the whole of the Columbia River basin, but the new proposal was limited to the East Side.

Thomas and Dombeck "sent back messages to the politicos that, because the primary environmental issue was going to be anadromous fish, we needed to extend our efforts to the entire Columbia basin. And so it was done." The basin assessment, expected to require two years to complete, was still under way eight years later during the 2001 interview. Even though Congress complained about the delays, "the powers in Congress are the reason that the process has gone on so long."

A potential existed for the "ultimate train wreck," Thomas believed, because the need for more water to sustain salmon habitat conflicted with the need to use the water instead to generate electricity and irrigate cropland, and because timber and grazing practices could cause siltation. Then there were Indian treaties that promised historic quantities of salmon. Add to that the major federal players—Bonneville Power Administration, Army Corps of Engineers, National Marine Fisheries Service, Fish and Wildlife Service, Forest Service, and Bureau of Land Management—that "ostensibly make the decisions for the management of the Columbia River." The process was working poorly: "the salmon are slowly sliding toward extinction because these various government entities simply can't come to grips with the issue."

FOREST FIRE AND TIMBER SALVAGE

"The phone rang in the middle of the night," Thomas recalled. "Deputy Chief Lamar Beasley said the first reports suggested we might have lost as many as forty people on Storm King Mountain in Colorado." It turned out that the fatalities numbered fourteen, but by the end of the fire season thirty-four had lost their lives fighting wildland fires.

Prineville Hotshots, Storm King Mountain fire, Colorado, 1994. Fourteen of this crew would be killed on that fire. David Frey photo.

Thomas and Dombeck were in Colorado the next day; it was a BLM fire but the fatalities were from a Forest Service "hotshot" crew from Prineville, Oregon.

They joined the surviving crew members, who "were on the hotel's patio drinking beer and holding something of a wake." Thomas gave his credit card to the bartender to buy a round. The firefighters "were stunned and scared to death that they or their supervisors were going to be scapegoated for the tragedy," and Thomas assured them that it would not happen. They were "tired of talking to headshrinkers and just wanted to go home." Thomas issued an order to let them go home the next day via a chartered airplane. He also directed that the "bodies that were going home were to be accompanied by a regional forester."

The single survivor of the blowup was hospitalized in serious condition. His parents were there and wanted to take him home to Seattle, but the trip would require a pressurized ambulance jet with an accompanying medical crew. An aide told Thomas that there was no authority to cover the high cost. Thomas asked the aide for a notebook and wrote a statement that he "lacked authority" to issue such an order, "but that I was doing it anyway." The aide said, "All right, but you may end up paying for this." However, there were no questions asked.

One aftermath of the fire season with its many fatalities was to "bear down" on safety, such as reducing night duty with its impaired visibility. "We were dealing with a philosophy of 'safety first—every fire, every time.'" Thomas believed that in some cases, fire crews had been trained to be "too aggressive." Thomas told forest supervisors in training sessions "to obey all the firefighting orders and lookout signals, but that they were in charge. 'If the hair is sticking up on the back of your neck, you get your people the hell out of there, and I'll back you.' I think it paid off."

Thomas thought that the Occupational Safety and Health Administration, which reviewed the Storm King fire, nearly always made judgments that the fault lay with whoever was in charge. "To them there is no such thing as an unpredictable phenomenon." In sum, OSHA "gets paid to investigate and assign blame." He acknowledged that mistakes had been made, and he was not all that upset about OSHA's conclusion that "there were problems and errors. I was quite upset with what I considered to be an outrageous insinuation that we didn't care."

The Forest Service, Thomas believed, needed to become "a fire management organization—not a firefighting organization." Fire responsibility should not be "collateral duty to something else"; there should be full-time fire managers. When there were many fires, the cadre would fight fires, but during years when fires were relatively few, they would "do controlled burns, and educate people and work with homeowners and agencies." Also, "we need to quit pretending" that fires are an act of God and to address the agency's traditional "collusion" with Congress on fire budgeting. The long-standing "game" played by the Forest Service and Congress assumes no fires and thus no money appropriated. Then, when fires occur, Congress passes "emergency" appropriations to pay the cost. Instead, Thomas believed, there needed to be a "realistic fire management budget."

Shortcomings aside, the Forest Service's historical fire prevention and suppression record is admirable indeed. In fact, recent critics insist that the agency has been too successful, since in many areas fire is a natural part of ecosystem development, and to exclude it is to create unnatural and "unhealthy" forests. Fire exclusion, then, has enabled fuel to accumulate that would have otherwise burned during smaller "natural" fires caused by lightning and in that way increased the likelihood of severe fires. Too, wildlife habitat is no longer "natural," and changes have had an impact on populations.

The topic of forest health led Thomas into a discussion of the timber salvage rider. The Rescission Act of 1995 was a lengthy law that made "emergency supplemental appropriations" available for a wide variety of purposes. Attached to the bill was a six-page section entitled "Emergency Salvage Timber Sale Program." The emergency was the unhealthy state of forests.

The salvage rider, as it was generally called, was very controversial in part because Congress had exempted the Forest Service from environmental requirements, such as consulting with the Fish and Wildlife Service on threatened and endangered species. Thomas ordered, nonetheless, that such consultations and other environmental regulations be followed.

Earlier, Thomas had stopped by Undersecretary Lyons's office, where he was watching television coverage of the House debate on the rider. He told Lyons that "we didn't really want this to happen, that it would be a political disaster." The rider passed by a lopsided bipartisan vote. When President Clinton signed the larger bill into law, he noted his reservations about the salvage rider. However, "the administration was pushing us hard to get the salvage out." Then "the administration changed its mind and started pulling back and buying back timber sales." The reversal had been "very astutely" orchestrated by environmentalists, who were able to get White House support. The Forest Service "got to be the scapegoat for a stupid decision in which we had no part."

The salvage rider had two parts: "one directed the intensive salvage of insect-infested, diseased, damaged, or burned timber." This approach fit well with the notion of treating unhealthy forests. But the other part of the rider, releasing "318 sales," had nothing to do with salvage and quickly became the more controversial.

Earlier, while Thomas and the Interagency Scientific Committee were studying the spotted owl, the substantial impact on the timber industry in Washington and Oregon, and perhaps California, became obvious. In response, Section 318 of the 1990 Appropriations Act for Interior and Related Agencies instructed the Forest Service to make two old-growth timber sales in each national forest that was within the range of the owl. "Most of the 318 sales were harvested in short order, but some of them were not cut and went into limbo over time." The salvage rider said that the in-limbo sales "would be released to the original buyers under the original conditions." The Forest Service was able to trade out of some of the sales—"we did everything we could to spare these areas"—but some of them were cut. "The enviros yelled, 'See what the Forest Service is doing under the guise of salvage—they are cutting old growth forests!' Salvage and the 318 sales were two very different things, and they damn well knew it. But, boy, they used the circumstance most adroitly as a propaganda tool. The Forest Service was the big loser."

The timber industry was also critical of the Forest Service because "we didn't do more." Even though Congress had exempted the sales from certain constraints, Thomas ordered that staff assume there was no exemption. "If we had not consulted with the Fish and Wildlife Service or other agencies, and if we had not done environmental assessments, we could have salvaged more volume." To Thomas, to have done so "wouldn't have been right—ethically or ecologically. We did what we thought we could do in an ecologically sound manner." Thus, the Forest Service "got whacked by both extremes on the issue."

In December 1996, Thomas retired after thirty-one years of federal service and accepted the position as Boone and Crockett professor of wildlife conservation at the University of Montana. He is also frequently invited to speak before various groups.

CHIEF MICHAEL P. DOMBECK

USDA FOREST SERVICE PHOTO

Michael P. Dombeck was able to ride out the spotted owl controversy while with the Bureau of Land Management. As Forest Service chief, however, he confronted equally vexing issues, from wildfire and grazing to road building and county funding. Building morale within the agency and bridges to its external critics was another focus for Dombeck.

He was born in Stevens Point, Wisconsin, on September 21, 1948, and graduated from the University of Wisconsin–Stevens Point in 1971 with a bachelor of science degree in biology and general science. He completed a master of science degree in biology and education in 1974 at Stevens Point and a second master's degree in zoology in 1976 at the University of Minnesota. In 1977, he learned that the Forest Service was looking for biologists. He applied and was selected, reporting for work as a GS-6 biological technician on the Hiawatha National Forest. It was there that he met his first "combat biologist," a label that reflected the difficulty that the traditional Forest Service culture had in accepting new disciplines. Biologists were considered "kind of a frill."

Dombeck did well, and he arranged to be assigned to a research project—as part of his job—that was central to his earning a Ph.D. in fisheries biology at

Source: An Interview with Michael P. Dombeck, *by Harold K. Steen. Durham, North Carolina: Forest History Society, 2004. Dombeck was interviewed in 2003.*

Iowa State University in 1984. He produced five peer-reviewed papers from that research and was active in professional societies, which provided him with national visibility, especially within the agency. As a result, he was promoted and moved to San Francisco as regional fisheries program manager for the California region. However, he quickly learned that the technical expertise that had served him so well in the Midwest no longer applied; the western species were different, the habitats were different, and the issues were different. Thus, he turned to developing long-range strategies and securing outside funding, efforts that added to his national prominence. In 1987, he was again promoted, this time to national fisheries program manager in Washington, D.C.

An early assignment was to promote a fishery program, Rise to the Future. He met and briefed Associate Chief Dale Robertson on the new program the same day that it was announced Robertson would succeed Chief Max Peterson. Robertson said that he wanted the program to be a "curve-bending event," something positive to counter the harsh controversies over the spotted owl and old-growth timber. Later, when Cy Jamison, director of the Bureau of Land Management, told Robertson that he wanted to bolster his agency's fish and wildlife management capabilities, the chief said, "I think I know just the person you need." With Robertson's assurance that he would always be welcome back in the Forest Service, in 1989 Dombeck transferred to BLM as science adviser and special assistant to the director.

BUREAU OF LAND MANAGEMENT

His BLM assignment went well, and by 1993, when President Clinton assumed office, the fish and wildlife program had reached about forty-eight million dollars toward a fifty-million-dollar goal. At the very end of the Bush administration, Dombeck, by now a member of the Senior Executive Service, had been asked to be acting assistant secretary of the Interior for lands and mineral management, an appointment that carried over into the early months of the Clinton administration. Interior Secretary Bruce Babbitt was one of the earliest cabinet members to be confirmed, and Dombeck worked with him during the early transition period. BLM Director Jamison, of course, had left with other Bush appointees, and when newly appointed Director Jim Baca was asked to leave because of his abrasive management style, Babbitt wanted Dombeck to "go down to BLM and settle things down for a while."

The secretary sent Dombeck's name to the White House, asking that the president nominate him as BLM director for Senate confirmation. It was now February 1994, a full year into the Clinton administration, during which time

Secretary of the Interior Bruce Babbitt and Bureau of Land Management Acting Director Dombeck during National Fishing Week, Washington, D.C., 1994. On January 6, 1997, Dombeck would be named Forest Service chief. Photo courtesy of Michael P. Dombeck.

Secretary Babbitt had begun an aggressive range management reform effort that Dombeck had supported. Senators, angered by this "War on the West," declined to act on Dombeck's nomination, and he was to remain as acting BLM director for three years. He believed that the interim nature of his appointment did not hamper him within the agency's culture, but it did cause him to focus more on the short term, rather than the long term.

GRAZING

It was New Mexico Senator Pete Domenici who "so cleverly labeled Bruce Babbitt's grazing policy as War on the West." Dombeck felt that it was ironic for the secretary to be positioned in that way, since his family was in the ranching business. He also believed that President Clinton "blinked" on the grazing issue, when key senators "got to the president early on." The presidential blink "really took the steam out of the boiler for Secretary Babbitt," and then it became a "series of compromises and long protracted meetings." In the House particularly, "we had tenacious hearings on issues where I would be sitting next to Jack Thomas and sometimes other agency heads in front of the House Resources Committee, just being drilled over this grazing debate."

The proposal to increase grazing fees prompted a "tremendous debate" because opponents said the higher fees "would put ranchers out of business." When Dombeck toured South Dakota, he learned that grazing fees on state lands were "anywhere from seven-fifty to eighteen dollars an animal unit month," plus the lessee had to pay the equivalent of property taxes. Then he stopped off at certain BLM lands where the agency had authority to auction off grazing fees. "That year we got something like eighteen dollars an animal unit month." In contrast, regular BLM "grazing fees were a dollar thirty-four." Later, Dombeck told Thomas and Babbitt that instead of fighting the Forest Service and BLM on grazing fees, ranchers "really ought to be throwing a big party every time we show up because they are getting such a good deal."

If the stockmen and their political representatives were fighting BLM on grazing issues, the environmental community did little to support reforms. Following the 1994 election, which gave Republicans control of Congress, Dombeck sat with Babbitt in the secretary's office. "He and I were on his speakerphone calling members of the environmental community, basically saying how much energy will your organization put forward to help us with this issue." Most responses were not supportive: grazing was not "their highest priority." Babbitt and Dombeck were disappointed, "knowing that we were going to have to backtrack even more." The lack of support was not expected. "Here was their champion in the Clinton administration, going up to Congress and crawling out, all beat up," and the secretary was also "being kicked" by the environmental community.

"After much dialogue, the grazing advisory board concept was deemed a good thing because ultimately you had community-based governance" for BLM lands. These local boards comprised representatives of the grazing industry, Indian tribes, county commissioners, the governor's office, and environmental groups. "They all worked very well, except in the state of Wyoming." There, the sole advisory council for the entire state was "very politicized." Even in states like Idaho that "were strongly antifed, this process worked and brought the government closer to the people."

Dombeck could see that in the Forest Service, the grazing issue was on a rougher track. "Jack wasn't getting much staff help" because the staff was trying to distance itself from the controversy. "They knew it would be a fight, and they were watching the spotted owl and all this other stuff, and the last thing they needed was another battle." Dombeck remembered going over to Chief Thomas's condo one Sunday morning, and "he and I went through the Domenici grazing bill line by line, doing an analysis of it." Dombeck took notes and typed them up. "It struck me as odd that in this huge organization, the staff wouldn't have done this kind of analysis for the chief."

Despite the controversy, Dombeck stated that "western ranchers are some of the nicest people I ever met. You can be on totally different pages on policy or philosophy, but yet they welcome you to their home and treat you just like one of the family." Ranchers are concerned that they will be "pushed out of the scene like the buffalo." He saw that their way of life was "just sort of slipping away. In reality, we should have never had grazing on ephemeral rangelands to begin with," where the land could not support grazing during times of drought. He continued, "In time, people's livelihoods and their financial positions, their ability to get loans linked to a federal permit and ephemeral rangelands is one of the unfortunate things that evolved in the U.S."

FOREST FIRES

"Right from day one, as I moved into the director's chair of the Bureau of Land Management, fire was really a tremendously traumatic issue to deal with." The worst fire, because of the many fatalities, was at Storm King Mountain in Colorado. "I can distinctly remember being on the phone with Jack Thomas several times that first night. We knew people had died. We didn't know exactly who they worked for and we didn't know exactly how many." They decided to fly out on the first available flight the next morning "to talk to employees." Dombeck thought it was strange that despite the many decades of fire suppression experience, the agencies had no formal policy on what to do when there are "a bunch of fatalities."

After they arrived at the scene, Dombeck and Thomas first visited with an injured firefighter in the local hospital. Then they went to the Holiday Inn in Grand Junction, where the crew was gathered. The chief and the director were in uniform, and Dombeck observed to Thomas that "there'll be a handful of these firefighters that know who you are, nobody's going to know who I am, nor will they care. Well, I was wrong." As did Thomas, Dombeck experienced "a tremendously emotional rollercoaster in talking to forty or fifty employees, some that had had a couple of beers too many already, some that were crying, some that were laughing, some that were embracing, some that were talking about what happened, and some that couldn't face it."

"Jack and I decided we needed to buy a round of whatever they were drinking and just started milling around and talking to people." The two leaders talked to every individual. "The psychological counselors were there, and the firefighters said, 'Get these guys out of our face. We don't need shrinks around.'"

In the midst of all this emotion, a smokejumper approached Dombeck and asked if he was BLM director. Dombeck said yes, and the smokejumper said that he was from Idaho and that his family were miners. "I'd like to talk to you about grazing fees." Dombeck's first thought was, "Wow, this is going to be a rough night." Instead, the jumper stated that he didn't think "it was fair that miners don't have to pay a royalty, while ranchers have to pay grazing fees." Later at the airport, while waiting for his flight home, Dombeck saw the jumper again. "'Mike,' he said, 'I just want to tell you how much we appreciate your coming out here,' and he took his smokejumper shirt off and handed it to me. That was one of the most touching things I ever had happen to me."

CHIEF OF THE FOREST SERVICE

Dale Robertson could be seen as a prototype for the agency's new fast-track career model, spending only a short time at each level until he reached the top. Dombeck, too, moved up rapidly, but on the staff side; that is, he had little managerial experience in the series of assignments that focused on fisheries. When he moved to BLM as science adviser and special assistant to the director, no one could have predicted that an unfortunate choice of BLM director at the change of presidential administration would have created an opportunity for Dombeck, age forty-five, to be named acting director—the first line assignment of his career. His inherent leadership ability came to the fore and made him visible to the politicos in the Department of Agriculture, who eventually invited him to return to the Forest Service as its chief at age forty-eight, just a year older than the "very young" Robertson had been.

As BLM director, Dombeck had opportunities to become acquainted with Assistant Secretary of Agriculture James Lyons, who was the direct political boss of the Forest Service chief. Typically they shared a podium at the National Cattlemen's Beef Association meeting and similar gatherings. In his presentations, "Jim was usually critical of the BLM and positive about the Forest Service. It's sort of the tradition at the Forest Service to view the BLM as the weak sister." In spring 1996, Lyons invited Dombeck to lunch to sound him out on becoming associate chief of the Forest Service when David Unger retired. Dombeck declined; he could see no rationale for leaving the top job in BLM for the second job in the Forest Service. Too, "when I looked at the challenges at the Forest Service, I was better off where I was."

Dombeck and Thomas met frequently to discuss a range of issues. Shortly after his meeting with Lyons, Thomas told Dombeck, "As I look around the Forest Service, I'm not sure that this administration would find any of the top team acceptable for the chief's job. I'm going to the secretary with a list. Can I put your name on it?" Dombeck said that he wanted to think about it. Others, including former chief Max Peterson, talked to Dombeck about their concerns about the Forest Service future, with its strained relations with Congress and the administration, plus low morale in the agency.

In early summer, Secretary of Agriculture Dan Glickman invited Dombeck to meet with him to discuss the chief's job. With the secretary's top staff present, they talked about Thomas's frustrations, and Glickman said that the chief was considering leaving. Would Dombeck agree to replace him? Again, Dombeck declined to commit. After talking with Thomas and Unger, Dombeck urged Thomas to stay on until the end of Clinton's first term. However, in the fall, Dombeck was asked again if he would be chief when Thomas left. He had several more discussions, but he was never specifically offered the job, nor did he ever say that he would accept.

During the several interviews, the secretary's office had agreed to all of Dombeck's demands, such as having authority over all non–Senior Executive Service personnel decisions. As had Thomas, Dombeck asked the secretary to be more visible at the Forest Service, and on tough issues, the secretary and the chief would both be on the podium to make the announcement. When Thomas did resign in early December, the secretary's office interviewed several other candidates. The same day that the cabinet was sworn in for a second term, Dombeck was invited to be chief of the Forest Service. His first day on the job was January 6, 1997.

The new chief's first order of business was to mend fences. Relations were "strained with Congress, strained with the administration, strained with the undersecretary's office—every place you looked there was a fight." Also, "the agency had lost its mission." Some internal and external relations would remain strained, but he was able to very substantially lower the levels of antagonism, many times by hosting breakfasts in the chief's office for political appointees, senators, and representatives and their key staffers, and other agency heads. Generally, Dombeck's guests were willing to meet him more than halfway; they too wanted to end the strained relations.

THE CHIEF'S DAY

"Dale Robertson is exactly right," Dombeck said. "There is no typical day." The chief found little time to reflect on big issues or think about the philosophy of management and "where we are going as a nation. We don't spend nearly enough time doing that." Instead, "we spend way, way, way too much time on the minutia of day-to-day management, the urgent issue that's biting you." In sum, the chief's job "is like drinking out of a fire hose when you aren't thirsty."

One type of day would be riddled with interruptions. A congressman might call, or it could be the Council on Environmental Quality as surrogate for the White House, or the undersecretary's office asking to meet with the chief, "literally all at the same time. There's this constant pressure that they always want to see the chief, and yet you can only be in one place at one time."

Other days were a series of meetings. "I've got my calendar here. I'll read off what happened that day." At eight A.M. was the daily chief and staff meeting that included the deputy chiefs. At nine-thirty there was a signing of a memorandum of understanding with the National Park Service, BLM, and the Continental Divide Trail Alliance in the Old Executive Office Building. At ten-thirty in the same building Dombeck met with CEQ to discuss the roadless issue. Then a meeting with the secretary's staff to review a watershed restoration project in Oregon. Following lunch there was a one-thirty meeting to give the chief a heads-up on plans for the 2002 Winter Olympics, some of whose major venues were on national forests. The next meeting was with the agency's legislative staff to prepare a presentation for Montana Senator Conrad Burns, who was an opponent of the roadless policy. The final meeting, at four-thirty, was also about roads but in the broader context of transportation policy. "If you notice, there's not a lot of time to sit back or think on days like this," he concluded.

"The most challenging days were during the congressional hearing season." During the first year and a half of his tenure, Dombeck "wanted to do all of the hearings rather than have deputy chiefs do them." The agency's legislative staff did "a great job of preparing the chief and anyone else who testifies." Typically, Dombeck would be up by three-thirty A.M. to review briefing materials; "It's like studying for a Ph.D. prelim." After breakfast at six, he would leave for the office. By then, he would have a list of questions for his staff that needed clarification. The staff would also arrive early on testimony days. Following a last-minute briefing, the chief would polish his opening statement and then head to the Hill for the hearing, which was usually over by noon. If it was a major hearing, such as oversight or appropriations, he would "come back and kind of melt away. Usually what I did on those days is walk around and visit employees and return phone calls."

Overall, "I really reduced the number of pieces of correspondence that I signed. The tradition was that the chief signed almost everything." Instead, Dombeck wanted the actual author, such as the recreation staff director, to sign and take credit where credit was due. "It really diminishes the clout of the chief's signature when almost everything is signed by the chief." He felt that many employees did not read official mail because they had no way to sift the important from the routine, if the chief signed everything. Nonetheless, "we always had this backlog of correspondence," and the secretary's office was scolding, "You're behind on these issues." Dombeck did manage to "reduce a lot of the volume" that came up to the chief and associate chief's office, keeping more of it at the deputy chief level. "Yet the culture was such that everything traditionally came up to the top, and we changed some of that, and some people liked it and some didn't."

Dombeck thought that "democratization as a result of e-mail is interesting." Although the new technology resulted in "great efficiencies," there were downsides. For example, as a result of e-mail, "the chief's office is rarely the first to break news." It was a great challenge to "beat the leaks and the grapevine." E-mail also allows people to put their personal spin on events as they interpret what the chief said within the framework of their own biases. Effective internal communication remains a big challenge, and "the chief must stay on top of it."

A demanding job with its constantly changing priorities and schedules, often externally imposed, routinely disrupted the chief's private life. Family vacations would take place without him: the secretary had called him to a meeting, or there was a fire out west. Then there were the evening receptions in official Washington, where it would be noticed if the chief were absent. "You just go and make an appearance and visit with a few folks and leave as quickly as you can." But there were the perks of office, too: extensive travel to beautiful places on public lands, a reserved parking spot at headquarters that made the chief in great demand for carpools, and invitations to participate in historic moments. On December 13, 2000, the Dombecks attended a Christmas party at the vice president's home. That evening, Al Gore made his concession speech to George W. Bush following the contested presidential election. "It's just a tremendous honor," Dombeck said, "to be affiliated with the greatest country in the world and the leadership and the opportunities you have that are also part of the job."

CHIEFS' SUMMIT AT GREY TOWERS

"One of the things that I wanted to do early on was to spend some time with the former chiefs. I also wanted to do something symbolic to help connect to the rest of the organization." Because of his recent transfer from BLM, "there were those who saw me as an outsider." For the meeting, Dombeck selected Grey Towers, Gifford Pinchot's magnificent home in Milford, Pennsylvania, which the Pinchot family had given to the Forest Service decades earlier. The five chiefs—McGuire,

Chiefs' Summit at Grey Towers, 1997: front row, Chiefs Dombeck and McGuire; back row, Chiefs Peterson, Thomas, and Robertson. At the beginning of his tenure, Chief Dombeck invited the former chiefs to Grey Towers to advise him on how best to deal with the many controversies surrounding the agency. USDA Forest Service photo.

Peterson, Robertson, Thomas, and Dombeck—"had a wonderful couple of days..., and that tremendous environment of Grey Towers can be spiritual, especially when you're talking about forest issues."

With each of his predecessors, Dombeck discussed "what was the most overarching lesson you learned as chief." Another topic was what "can a chief do to rebuild credibility of the organization." The chiefs also discussed how to anticipate issues and "how can we be more proactive and prospective, because many people felt that we were in a totally reactive mode." Basically, he wanted to know how to be a better conservation leader.

"What I took home from that summit was the fact that there was complete agreement that we're not talking nearly enough about values, about the values of forests, about the value of conservation." Dombeck decided to "use the bully pulpit that comes with the chief's office...that really no other agency has." In particular, he "made a concerted effort with the communications staff...to turn the image of the Forest Service around in the press." When he became chief, the press had been about eighty percent negative "as a result of some debacles like the spotted owl and the salvage rider." The shift was dramatic; in a few years the press was eighty percent positive, thanks to "largely talking about values and about many of the positive things that the agency does." From his bully pulpit, Dombeck had emphasized such topics as "the importance of watershed functions, sustainable forestry, and ecosystem management." He observed that "we did rebuild some trust and take a hold of the agenda and set the agenda."

CONTRASTING AGENCY CULTURES

"The Forest Service has so much more organizational horsepower than any other land management agency. It certainly dwarfs the BLM." Dombeck believes that the Forest Service culture had worked to the agency's advantage until the 1980s, when "the power" began to move out of Washington, D.C., to local land advisory groups. The Forest Service had more difficulty adapting to that shift than did BLM. "I believe that the BLM is more decentralized in the way it operates"—an opinion that runs counter to the proud forest ranger's traditional view that the Forest Service is more decentralized, and less politicized, than agencies in the Department of the Interior. In the Forest Service, Dombeck said, it is the chief and the Washington office that the agency culture sees as the center of power, whereas in BLM, it is the several state directors' offices that have the clout.

The Forest Service position is that "we're the trained professionals, we're very good at what we do. Trust us, we know best. But the public wasn't buying it....From the standpoint of the way the employees behave and what's considered right, the Forest Service is much more rigid. There's a lot more kissing of the ring in the Forest Service." As a poorer agency, Dombeck believed, BLM was much more cost conscious, more prone to carpooling, showed more concern over expensive travel, and was able to transfer more funds from Washington to the field. "In the end, the BLM ends up being a lot more flexible and less labor intensive."

Then there is the very sensitive matter—to the Forest Service—of having a political appointee as agency head. Since the Federal Land and Policy Management Act was enacted in 1976, the BLM director had been appointed by the president, subject to Senate confirmation. Jack Ward Thomas's Schedule C, or political, appointment as chief was seen by the whole agency as a severe blow to its culture.

Dombeck readily agreed that not all political appointees might be fully qualified to head an agency and that membership in the Senior Executive Service ensured an impressive level of experience and skills. However, he said, the at-times fortresslike posture of the Forest Service with its "professional chief" could keep it out of the decision-making loop. He cited the spotted owl and New World Mine controversies as issues that created a great deal of tension between the Forest Service and the administration. In both cases, the Forest Service was not at the table when basic policy decisions were made, but the political appointees in the Department of the Interior were.

When Dombeck moved over to BLM to be science adviser and special assistant to the director, he elicited a promise from Director Cy Jamison that he would have no involvement in the spotted owl issue, a "debacle waiting to happen. I wasn't about to compromise my professional credibility over whatever direction we would end up going." But he did know that for a crucial meeting at the White House, Chief Robertson was out of the country, and Associate Chief George Leonard was not invited. "Here you had Cy Jamison with the BLM having some thirteen percent of the old-growth spotted owl habitat in the Pacific Northwest and running the show at the White House." Despite its extensive old-growth holdings, the Forest Service was absent while it was "essentially getting rolled."

The New World Mine, Dombeck's other example, is located on the Gallatin National Forest in Montana, near Yellowstone National Park. The mine had lain dormant for many decades, and when the owners proposed to resume operation, the park superintendent proclaimed that Yellowstone was in danger. "From my perspective as BLM director," Dombeck recalled, "this is the way I saw it. The White House wanted something as an environmental gesture that was meaningful." Through some process the New World Mine was selected; it would be bought out, to prevent its reopening. The Forest Service had objected to even having the mine on the list of options for consideration. Nonetheless, "the Department of the Interior opinion prevailed at the White House, and the decision was made that we're going ahead and buy this thing out." The Forest Service "could never accept the fact that it lost. It continued to fight that issue even though the orders were there from the president." Since the Forest Service "wasn't close enough to the White House to be at the table when the actual debate occurred, it probably wasn't aware of all of the factors." The agency would not stop fighting, and it "was one of the things that really strained the relationship between the administration and the Forest Service." It became viewed as this "unruly troublemaker that had to have its own way."

There were times, however, when the cultures were in sync, such as the Forest Service–BLM interchange that took place while Peterson was chief. "One of the best ideas that came out of the Reagan administration was the interchange, and it went nowhere." Dombeck continued, "The people that tanked the interchange were probably the employees of the agencies." At the field level, neither could support "losing acres" to the other, and that "local employees of each agency lobbied their local constituencies and county commissioners, who through their congressional delegations opted to maintain the status quo." This grassroots activity was easy to accommodate "because no congressional committee wants to give up an inch of oversight authority." He hoped that someday the "stars will line up" in the White House and Congress, and something akin to the interchange would happen.

WORKING WITH THE WHITE HOUSE

Before he was chief, Dombeck had talks with Council for Environmental Quality Director Katie McGinty; "they were always positive." Dombeck surmised that McGinty "had the confidence of both the president and the vice president" because no one questioned her authority to make decisions. But when George Frampton, the assistant secretary of the Interior for fish, wildlife, and parks, replaced McGinty at CEQ early in Clinton's second term, "the whole dynamic really changed. George always asked us how we felt about issues." The dynamic changed again when John Podesta became presidential chief of staff. "John and George both had a personal interest in our issues, and that's when the president started to take interest."

Dombeck discovered that "when the president and the White House get involved in an issue, a lot of the coordination problems and bureaucratic stalls just sort of fall away." White House support explains how Dombeck had been able to produce a roadless rule "too fast," as some had complained. He compared the speed of the roadless rule to that of the Northwest Forest Plan for the spotted owl in 1993, when several agencies, under presidential orders, had completed the plan in months rather than years—or not at all—as would have happened if the agencies had been working without such a mandate.

"Sometimes I had meetings in the chief of staff's office three times a week, usually without Dan Glickman." Because he had the secretary's confidence, as well as that of Undersecretary Lyons and others in the administration, "we were really able to do a lot of things." Dombeck "always found the meetings with Frampton and Podesta to be very, very cordial. The dialogue usually began, "How would the Forest Service like to proceed?" There was not always agreement, however, and the chief often found himself pushed "further toward the green side than I was willing to go."

ROADLESS AREAS

Secretary Glickman invited Dombeck to attend USDA leadership meetings. The chief attended perhaps one per month, and after one of those meetings he was chatting with Greg Frazier, the secretary's chief of staff. Glickman stopped and listened in while Dombeck was explaining that the agency still lacked a strategy to deal with tough issues. The secretary said, "Mike, I want you to develop a strategy, something that makes sense that we can go with." Out of that ad hoc discussion grew the Natural Resource Agenda, which the Forest Service developed to articulate its goals. The agenda included recreation, watershed health and restoration, sustainable forest management, and roads. "That began to turn the ball game around, and the Forest Service then began to take the lead on issues. And I have this theory in Washington, if you're dealing with controversial issues, it's either you keep them busy or they'll keep you busy."

The roads issue was especially tough. "A lot of people didn't realize that we had a letter from about a third of the members of the House of Representatives saying, 'Stop building roads in roadless areas.'" In fact, there had been a steep decline in the agency's road budget, and as a result, "less than twenty percent of our roads were being maintained to environmental safety standards." It was clear that the debate over the road budget, like that over the spotted owl, was "really a surrogate for 'Let's not cut any more old-growth.'" The road budget had been cut to prevent building roads into roadless areas that contained old-growth timber.

Dombeck phoned Idaho Senator Larry Craig to say he was about to announce a moratorium on road building in roadless areas. "Obviously, he did his best to talk me out of it," saying, "Mike, don't do that. If you draw that line in the sand, we'll never get beyond it." But Dombeck did, and "the explosion was fairly intense from the opposition." The tactical mistake was not involving the field more, "in letting the forest supervisors know exactly what was going on. They were getting calls from the press, and they didn't know what to say." Dombeck had dealt with a tough issue by acting quickly; a long discussion would have allowed "the extremes" to take the debate away from the Forest Service. "The challenge in a large organization is how do you include enough people so the issue doesn't get away from you." The moratorium also surprised Congress "that the Forest Service would make such a bold move in a major national policy issue." At the time, Dombeck thought that if the roads-in-roadless ban prompted Congress to increase the roads budget from ninety million to four or five hundred million dollars, "then all this

President Clinton announcing the roadless rule in a ceremony at the National Arboretum, with Secretary of Agriculture Dan Glickman, Chief Dombeck, Senator Gaylord Nelson, and EPA Administrator Carol Browner looking on, 2001. The ban on roads in roadless areas was one of the major controversies of Dombeck's tenure. Photo courtesy of Michael P. Dombeck.

will have been worth it." He later noted that, since then, the roads budget had turned around and had indeed been increasing rather substantially. But at the time of the announcement, "some forest supervisors called up and said, 'Hey, Chief, you've really stirred up a hornet's nest.' Other forest supervisors said, 'Thank you, Chief, for getting this monkey off my back.'"

Dombeck had talked with Secretary Glickman's chief of staff, Greg Frazier, three times about the pending roads-in-roadless controversy and told him, "This is going to be fasten-your-seatbelt time when we do this." He wanted to be certain that the secretary was fully aware "that we were going to move into some fairly intense debate. When the administration saw the support for the moratorium on road building in roadless, they became more involved. The environmental community became more involved, and the thing unfolded with the president asking the Forest Service to take a look at roadless and develop a policy." In January 2001, President Clinton, in a ceremony at the National Arboretum, announced a roadless rule that made Dombeck's moratorium official administration policy.

"Much of the opposition to the roadless policy," Dombeck believed, "was from people that didn't want more land available out there for wilderness designation, and if they could punch a road into this, then it wouldn't qualify."

Some foresters complained that wilderness designation took away their ability to manage. Feelings were so intense within the forestry community that the Society of American Foresters notified Dombeck that the outcome of its ethics investigation would determine whether he would be stripped of his membership for unethical behavior in the roadless controversy. The organization's investigation uncovered no ethical violations.

ALASKA

"From my first week as chief, probably about half of my time was spent on the Tongass National Forest." The Tongass was "another example of a potato so hot that staff was afraid to deal with it, and hence it ended up on the chief's desk almost immediately." Dombeck continued, "My first day on the job was Monday and by Wednesday I felt like I had my feet in concrete with the weight of the Tongass issue and the various negotiations." Central to the tension was the Alaska congressional delegation—Senators Murkowski and Stevens and Representative Young—"literally half of the oversight in appropriations process of the Forest Service." The delegation "brought the regional forester and others from Alaska back to Washington over twenty-five times for hearings." Dombeck felt that quieter hearings held in Alaska, away from the national press, would have yielded better results.

"We did make forest management mistakes in Alaska," some driven by congressional pressure on long-term timber sale contracts, roads, and below-cost timber sales. "If there ever was a government subsidy, it was certainly in Alaska. It probably would have been better and a lot cheaper just to send them the money. Alaska was a big challenge." The dichotomy of opinion—many Americans consider Alaska our last source of frontier and wilderness values, and many Alaskans contend that Alaska's resources are Alaskans' to develop—was a source of tension common throughout the American West as well. However, the buying out of the long-term contracts, plus the ban on new roads in roadless areas, reduced the level of controversy.

THE TWENTY-FIVE PERCENT FUND

"Since the time that Pinchot was chief, the Forest Service has returned twenty-five percent of timber sale receipts to the counties." The basic rationale for this transfer of funds was that federal lands paid no local taxes. Especially in counties that contained large federal holdings, the county tax base would be adversely affected. Thus, these payments were "in lieu of taxes." Times change, and Dombeck said that "the first three or four months I was chief, we had about twenty-five million dollars in timber sales enjoined in Texas. So that's about five, six million bucks that school systems were not getting because of the litigation." This meant that base support for education was linked to a judge's decision. "You've really got to ask the question that in this day and age, should funding for an important social benefit like education or roads be tied to one of the most controversial programs in the Forest Service?" Even though in one sense the agency's timber program had a built-in constituency of local teachers' unions, school superintendents, parents, and anyone else who saw education as important, the increasingly intense controversy over logging and road building suggested that change was in order—that the twenty-five percent fund had outlived its usefulness.

An important precedent had been set by the Northwest Forest Plan for the spotted owl, which provided federal relief for counties affected by a reduction in logging: the counties would continue to receive the same level of revenue for ten years. However, when Dombeck began discussions on changing the fund, he heard that it would be unlikely to find support within the agency. "They thought I would run into a buzz saw." He continued to pursue the idea and talked with some members of Congress and people in the administration.

Dombeck was a speaker at a National Association of Counties meeting in Washington, D.C. "I walked into the room and there sits Idaho Senator Larry Craig; he had invited himself." After an amiable conversation about the twenty-five percent fund during lunch, the two spoke about it to the group of county representatives. Craig referred to Dombeck's proposal by stating, "This dog won't hunt." The senator went on to say, in effect, that it was a "dumb idea and we've got to beat it back." When Dombeck spoke, "There probably weren't a handful of people in the room that agreed with me." He continued, "But we got the administration to support it."

An important turning point was when Oregon Senator Ron Wyden became interested. He was the ranking Democrat on the subcommittee that Craig chaired. "Wyden was up for reelection in 2004, and the spotted owl safety net expires about that time for the Oregon counties." Clearly, loss of the owl money was the "last issue that he wants to go into a campaign with." Negotiations continued, and "finally the third year we got Craig to cosponsor the legislation with Wyden"; the bill passed and was signed into law. "That was the only time I was invited to the White House to the Oval Office to witness the signing of a piece of legislation."

Changing the twenty-five percent fund "was one accomplishment that I am proud of that I hope will make a big difference in the long haul," Dombeck said.

"We made some compromises, but that is what legislation is all about." One of the compromises was to give the affected counties a choice between continuing with the old system or going with the new system. Under the new system, the counties would receive payments based upon "the average of the highest three years for a ten-year period." That made the counties speculate; if timber receipts went up, the old system was to their advantage. If receipts went down, the new system would be better for them. The law was a definite improvement, but Dombeck believed that "Congress could get the agencies and counties beyond all this red tape by simply legislating a fair payment in lieu of taxes program for all public land."

THE PRESIDENT TOURS A FOREST FIRE

"It was Friday night. I'd probably been out West for about two weeks on fires, and the phone rang. It was somebody from the secretary's office saying that we think maybe the secretary or somebody from the White House may be coming out on the fire. I thought, Gee, I thought the secretary was in Europe." Dombeck was told to check back in the morning before he caught his return flight to Washington. The next morning he was told to go to Boise, that someone from the White House might be coming, and Dombeck was to help with the advance. The chief guessed that it was the president, and then he finally got official word that indeed President Clinton wanted to see a fire, and what were his options? Given the especially tough fire situation during the summer of 2000, there were ready options in California, Nevada, Montana, and Idaho.

"This was August and this was a presidential election year." Because California had fifty-four electoral votes, it was a "no-brainer" that the president would pick a fire in California. Instead, "the word comes back that he wants to come to Idaho, the state where there probably wasn't a handful of people in a hundred-mile radius that had voted for him, and the congressional delegation was completely opposed to him. In fact, the senior senator disliked Clinton so much that he didn't go to the State of the Union two years in a row." Later, when Air Force One landed at the Boise airport, who walks off behind the president but Senator Larry Craig," the very senator who had shunned the State of the Union. The group transferred to a smaller jet that was temporarily renamed Air Force One, as any plane the president flies on carries that name. On board were the president, Secretary Glickman, Secretary Babbitt, Idaho Governor Dirk Kempthorne, Senator Craig, presidential Chief of Staff Podesta, and Chief Dombeck.

Chief Dombeck visiting a fire crew, Montana, 1999. The following summer, President Clinton toured Idaho fires with Dombeck. USDA Forest Service photo.

"We knew that Craig and the governor were going to push for more salvage," with all of the newly fire-killed timber available. Coincidentally, the Lewiston paper that very morning had carried an article about local mill closures due to a depressed market. The article was solid evidence that the local markets could not absorb the salvaged logs. Dombeck had a copy of the paper but was reluctant to hand it to the president himself, "so I gave the paper to Bruce Babbitt. When the senator brought up the issue of salvage, Babbitt handed the paper to the president with a smile on his face. The president handed the paper around, and that ended our salvage discussion." Later, as they toured the fires via Marine One, the presidential helicopter, Clinton "asked a lot of questions about the fire and then had a wonderful meeting with the employees and firefighters that were on the line that were gathered for the ceremony." During the flight back to Washington, "we really had a very productive discussion with Senator Craig about where to go with fire."

THE FIRE BUDGET

"When the president goes on a trip like that, obviously he's got to do something. The mandate that he gave the two secretaries was to come up with a strategy and request for the resources needed to deal with the fire issue." In response, the Forest Service, Bureau of Land Management, National Park Service, Fish and Wildlife

Service, and Bureau of Indian Affairs developed an interagency fire plan, working closely with western governors. The effort resulted in "perhaps the biggest budget increase in the history of the Forest Service," with the agency receiving 1.4 billion dollars out of a 2.2-billion-dollar fire-related appropriation for the several agencies. Dombeck thought it ironic, since only three years earlier—in reaction to the road-building moratorium—Alaska Representative Young had warned that Congress was going to cut the Forest Service budget "'until you squeal,' or more appropriately, they were going to bring us back to custodial management." At that time, no one would have predicted any increase, let alone one of such magnitude. "I think it's a tremendous advantage when you can have presidential involvement, because whether it is the Northwest Forest Plan, the roadless policy, or the fire issue, without that, none of these things would have really happened to the extent and level that they did," Dombeck said.

"The bulk of that appropriation went basically to rebuild the firefighting machine of the agencies—the suppression effort—because we were at less than seventy percent, maybe sixty, of MEL, which is the 'most efficient level.'" In addition to improved suppression capability, Dombeck believed that "the Forest Service now has tremendous opportunities in moving forward with fuel reduction and focusing on the condition we want the land in." He gave priority to urban-wildland interfaces, "to really do the work on the land in the areas that are going to make a difference." He hoped that subsequent administrations would not give emphasis to timber harvest as part of the fuel reduction effort, "so that we don't replay the fights of the salvage-rider era."

WORKFORCE DIVERSITY

A big issue that remained challenging throughout Dombeck's tenure as chief was workforce diversity. He inherited a "major backlog of complaints. Some complaints had been on the books for ten years or more. It was truly a bureaucracy run amok." The backlog contained a "thousand plus" complaints.

Dombeck made a "whirlwind tour of the Forest Service, hitting every station, region, and the state and private areas." His aim was to get the organization focused on the importance of civil rights by using "the symbolism of the chief doing this tour." Then he asked Kathy Gugulis from the Natural Resources Conservation Service to head up a team that included Forest Service staff "to really clean up the backlog issue. She structured it just like an incident command team would on a fire, and I'm proud to say they really did eliminate that backlog."

The chief "met routinely with the various representatives of the Hispanic community, the African American community, and the Native American community."

As with his national tour, he wanted all employees to understand that civil rights and workforce diversity were a priority. As to recruiting from outside the agency, Dombeck pointed to the selection of Hilda Diaz-Soltero as the first female associate chief, the second-in-command for the agency. Diaz-Soltero had been the secretary of natural resources in Puerto Rico, a cabinet-level position. She also had experience with the Fish and Wildlife Service and the National Marine Fisheries Service. Her "tremendous project management skills and tremendous energy" were additional abilities. "Some would say you're just playing favorites to get people of color or to get women in top jobs. The fact is, I don't think we made any compromises. We really did get some good people."

TRANSITION FROM CLINTON TO BUSH

As the Forest Service leaders considered the forthcoming 2000 presidential election, they went through "the what-if scenarios: what if Gore gets elected, what if Bush gets elected." Dombeck expected that if Al Gore was elected, the Forest Service would build on the road ban in roadless areas and "call a halt to commercial harvest of old-growth. We had already begun to make statements about that." They would also revamp the "outdated timber sale contract approach to vegetation management." Since the vice president had shown little interest in Forest Service affairs, Dombeck did not expect him to be particularly visible in the agency's programs after the election. It was more likely that Gore would be involved with high technology, fuel cells, energy efficiency, and so forth.

"With George Bush, we really didn't know where he was coming from." During the campaign, Governor Bush had stated his opposition to the road ban, which some western governors would urge him to overturn. Beyond that, Dombeck assumed that Bush's conservation policies would be "more in line with his father's." After the election, Dombeck worked with Bush's transition team, recommending that they look at the road ban within the context of the largest budget increase in agency history—to overturn it could trigger widespread congressional opposition.

As soon as Ann Veneman was nominated as secretary of Agriculture, Dombeck wrote offering to brief her on the major issues in preparation for her confirmation hearing. He believed that Veneman "didn't have much decision space, that the mandates for where we were going in natural resources were really coming from the transition team through the vice president's office." It quickly became clear to Dombeck that the new administration wanted him out, "and the next step was basically to have this transition in a way that it would be most benign for the organization. What we learned in the transition between Dale and Jack, that's the last thing the agency needed." At all times, Secretary Veneman was "very, very, gracious."

Veneman was so gracious that Dombeck was almost afraid she was going to ask him to say on as chief. But she didn't, instead asking him to prepare a list of "people that you think ought to be the next chief." There was time to choose, since under civil service regulations, Dombeck could remain as chief for one-hundred twenty days following the presidential inauguration. However, he resigned effective March 31, 2001.

The list that Dombeck prepared for Veneman contained seven names, "all in the Forest Service leadership." The secretary asked about some of the names, and Dombeck responded with pros and cons of each. A member of the secretary's staff pointed to one name, saying that she "probably wasn't an acceptable candidate" because of her open support for the road ban in roadless areas. The secretary and her staff "were very courteous and respecting of the chief and all employees of the Forest Service. They knew that they had to select a chief that was a resource professional."

Dale Bosworth, regional forester in Missoula, was on Dombeck's list. "He was in the top tier with regard to his land ethic. He had good people skills. He had a good presence. His temperament was appropriate to be able to handle the stress and controversy." And then a keen observation: "He had the support of the rank and file, certainly more than Jack and I did," an oblique reference to the agency's cultural resistance to outsiders. Bosworth became chief of the Forest Service on April 12, 2001.

On September 1, 2001, Dombeck began teaching at his alma mater in his hometown, the University of Wisconsin-Stevens Point. His full title is Professor of Global Environmental Management and University of Wisconsin System Fellow of Global Conservation. In addition to his academic responsibilities, he writes and speaks on conservation issues. Too, he fishes and canoes whenever he can. Despite his lofty title and impressive résumé, his voice-mail instructions for his university telephone are, "This is Mike. Leave a message."

AFTERWORD

The Art of the Possible

Government institutions evolve as society changes—demographically, economically, and socially. Gifford Pinchot knew that. In his famous maxims he wrote, "It is more trouble to consult the public than to ignore them, but that is what you are hired for. Find out in advance what the public will stand for. If it is right and they won't stand for it, postpone action and educate them."

Pinchot, of course, was always confident that he could sell his point of view to anyone. And he usually could.

Chief McGuire put it this way: "My attitude, and perhaps it is different than some, was not that the professional simply knows best, but rather an attitude that since these national forests are public property, the public can decide to manage them any way at all." It was important to find out what the public wants done "and then do it."

One of the most telling polls I ever saw was in the late 1970s, when a majority of the people of Oregon indicated for the first time ever that the existence of old-growth forests was more important to the state than timber-related jobs. It was a political sea change.

But what has this to do with how the chiefs of the Forest Service do business? The answer is, Everything, because in a democratic system they belong to everybody.

The chiefs belong first of all to the presidential administration, since the Forest Service is an executive branch agency. The chief answers to the Department of Agriculture's undersecretary for natural resources, who these days also oversees the Natural Resources Conservation Service. Next up the ladder are the secretary of Agriculture and his deputy. Above the secretary are the Office of Management and Budget, which sets the president's spending priorities, plus the Council of Economic Advisers, the Council on Environmental Quality, and other White House desks, such as the Office of Science and Technology Policy.

by James W. Giltmier

Former professional staff member, Senate Committee on Agriculture, Nutrition, and Forestry; former executive director and senior fellow, Pinchot Institute for Conservation; and former Washington representative, Tennessee Valley Authority.

The Department of Justice, the Environmental Protection Agency, the Department of Labor, the Department of Commerce, and several agencies of the Department of the Interior may want to meddle, too. At the top is the president, who seldom gets involved—unless his name is William J. Clinton or George W. Bush.

But the 535 members of the House and Senate know who really owns the Forest Service. They do, because Congress holds the purse strings.

Congress does not sit up on Capitol Hill chunking out one law after another. Congress spends a lot of time looking over the shoulders of the executive branch agencies to see whether the laws it enacted are being carried out in accordance with the desires of its constituencies, and it tries to put new shadings on existing laws that the original authors never dreamed of. The agencies, in response, dedicate a lot of their energy to avoiding the dictates that Congress has given them against their advice. The result is a democratic game that makes chess look like child's play.

The operation of a democratic government is about who has power and how it is used or abused, about basic civility and public service, about checks and balances, and yes, about outright scorn for the democratic process. Government can accomplish only what is possible at the moment when consensus is reached.

Any federal administrator who does not maintain an awareness of politics—as well as the fundamental changes occurring in the society—is leaving his employees and their mission vulnerable and unprotected. We learn in this book that Chief John McGuire spent half of his time gathering intelligence, somewhat less than that conducting external relations, and the remainder, motivating the staff. It is somewhat surprising that he did not mention time spent defending the institution of the Forest Service like a mother grizzly. McGuire wore that on his sleeve.

Chief Max Peterson did that, too. He was left to carry out the Forest and Rangeland Renewable Resources Planning Act—the RPA—in a hostile environment. The authorizing committees of Congress wanted the president's budget to reflect what the Forest Service budget ought to be based on the potential of the agency to follow good science and multiple-use management.

The appropriations committees mostly didn't care about that. Then as now, these committees were dominated by westerners who used the spending bills on public lands to distribute pork barrel projects to their states. And against congressional rules to the contrary, they legislated management of the public lands as "riders" to the spending bills.

Chief Peterson was audacious. Caught in a vise between the legislating (authorizing) committees and the appropriations committees, he brought to Capitol Hill two Forest Service budgets, one to satisfy the White House and the appropriations committees and one to show what the agency could accomplish for the nation if it had the money.

The Forest Service over the course of a century has transformed itself from forest custodian in the early years, to commodity provider during and after World War II, to wilderness provider and ecosystem manager today. It is—and always was—the agency that it is permitted to be each day, based on the myriad outside influences that move it in one direction or another.

The agency's problem has been that the public never could make up its mind what it wanted the Forest Service to be. The chiefs, dedicated public servants all, try to find out in advance what the public will stand for; anyone caught in the middle is deemed to be unsuccessful.

The seven chiefs whose voices we hear in this book have seen many changes. When the chief and all his deputies were located across the hall from one another in Agriculture's South Building, the proximity to each other and to the higher-ups in the department promoted an informal atmosphere. In 1990 most Forest Service employees in the Washington office were packed off to the Yates Building, or the Auditors Building, as it is more commonly known, across Fourteenth Street. The deputy chiefs have been scattered to the corner offices of the headquarters building, and the inevitable loss of contact and collegiality complicates decision making. And most of the professional talent has always been located in the woods, where they can do the most good—or harm, depending on your point of view.

Congress has changed as well. Each of these seven chiefs faced a different sort of legislative branch. There was a time, prior to the mid-1970s, when committee chairs were omnipotent and the freshman members of both houses of Congress were menials. New senators were generally not permitted to give their first speeches until given the O.K. signal by a veteran member. In the House, the Appropriations Committee, where all spending bills were born, Representative Jamie Whitten of Mississippi reigned as chairman for many years, and his subcommittee chairs, each parceling out the money for one or more agencies, were known as the College of Cardinals.

In both chambers arcane devices established in the rules could hurry legislation on its way or stop it cold, depending on the predilection of the leadership. In the House, the Rules Committee could stop any bill, no matter how popular. The Senate had its famous filibusters, among other devices, designed to make it difficult for the apparent popular will to prevail. A bystander might ask, "How could they do something like that?" The answer was, "Because they wanted to." While the House often acts like a churlish mob, for good or ill, the Senate uses its deliberative rules to keep silly things—or important things—from happening.

The congressional context of the first sixty-five years of the Forest Service actually made it somewhat easier for the early chiefs to do business on Capitol Hill. The same road blocks were there every day. The same people were standing in the same places every day, saying the same things. It might not mean happiness, but it meant stability. Furthermore, prior to the first Earth Day in 1970, the public was largely unaware of its ability to stir things up. Then suddenly hundreds of paid lobbyists were swarming about the Hill, and today there are thousands, representing nonprofit causes like the environment as well as less altruistic interests.

The dilution of power in Congress began in the House in the early 1970s and slowly affected the Senate, where one day a new senator from Florida announced the advent of "government in the sunshine." He had the door to his private office removed. One of the senior members of the Senate Committee on Agriculture and Forestry asked, "Who is this jerk?" He was the future.

Within a decade, there was a greater sharing of power in Congress. Chairmanship kings were demoted to satraps. Even Jamie Whitten had to share power. Americans began to bury their senators and representatives in mail and faxes and eventually e-mail, and suddenly chiefs had to worry not only about citizens' reaction to field decisions and appeals but also about case work generated by the congressional offices, where members had, almost overnight, become ombudsmen. Representatives and senators pursued these cases until their constituents were happy, even if the constituents were wrong. And as far as their elected representatives were concerned, they were always right.

When Richard McArdle was chief in 1952, America was happy to get the timber from the national forests and build itself out of the economic stagnation of the Depression and World War II. McArdle could count on both hands the people who played a major role in what it was possible for him to do. By the time Michael Dombeck resigned in 2001, there were national organizations dedicated to the idea that timber harvesting should cease forever in the national forests. Citizens were being trained in guerilla tactics to stop the chainsaws. Mail, lawsuits, and appeals had increased thousands of times over. Several other federal agencies had acquired a say-so about individual timber sales, including something called the God Squad. Local decisions previously made by forest rangers had to be kicked upstairs for the chief's signature. And perhaps worst of all, any member of Congress—which was polarized over how the public forests should be managed—could now exert influence on how the chief did business.

And whose fault was that?

It was an eighty-six-year-old lady in Boston.

During one of the 1970s Senate debates over the disposition of public domain lands in Alaska, she sent a letter to Senator Hubert H. Humphrey of Minnesota. "Dear Senator Humphrey," she began. "I hope that you will vote to save 800,000 acres of Alaska forests as wilderness. I have not been to Alaska yet, but I want to know that those trees are going to be there just the same." At the same time that letter arrived in Senator Humphrey's office, large groups of burly Alaska loggers were striding the halls of the Senate office buildings, offering Congress a different view of how Alaska's resources ought to be allocated.

Humphrey, like many in Congress, had high regard for lumberjacks, but he liked little old ladies even more.

Each of the chiefs has tried to influence the course of the Forest Service according to his own vision of the way things should be, but it's not really a tautology to say that things are the way they are because that's the only way they could be under the circumstances. That is why whatever the future holds, Pinchot's maxims will be framed and hanging on the wall.

INDEX